EFFECTIVE ADULT LEARNING

EFFECTIVE

with chapters by

Bryan L. Garton

Assistant Professor
Agricultural Education
University of Missouri–Columbia

Steven R. Harbstreit

Associate Professor
Agricultural Education
Kansas State University
Manhattan

W. Wade Miller

Professor
Agricultural Education and Studies
Iowa State University
Ames

ADULT LEARNING

by

Robert J. Birkenholz

Professor
Agricultural Education
University of Missouri–Columbia

Interstate Publishers, Inc.
Danville, Illinois

EFFECTIVE ADULT LEARNING

Library of Congress Catalog Card No. 98-72262

ISBN 0-8134-3160-3

1 2 3 4 5 6 7 8 9 10   04 03 02 01 00 99 98

Order from

Interstate Publishers, Inc.

510 North Vermilion Street
P.O. Box 50
Danville, IL 61834-0050

Phone: (800) 843-4774
Fax: (217) 446-9706
Email: info-ipp@IPPINC.com

World Wide Web: http://www.IPPINC.com

To Pam, Todd, *and* Reed, *with all my love.*

———————————

In memory of

Dr. Alan A. Kahler

FOREWORD

This manuscript was conceived and prepared to provide a comprehensive overview of the principles and practices involved in facilitating educational programs for adults. The wide-ranging topics addressed in the text provide foundational information and guidelines for present and future adult educators to consider as they contemplate decisions regarding adult education programs.

The text was not intended to serve as a reference for a graduate-level theory course on adult learning. Rather, the content and materials were organized as a practical reference for upper-level undergraduate or graduate students who were interested in planning and conducting educational programs for adult audiences. Current adult educators may also find the information and guidelines included in the text to provide useful ideas to improve their program offerings.

This volume represents the collective thoughts, ideas, and efforts of literally hundreds of colleagues who have contributed directly and indirectly to the contents of the pages that follow. For this reason, I gratefully acknowledge all others who knowingly or unknowingly contributed to this manuscript. Specific recognition is due to my colleagues, Professors Bryan Garton, Steve Harbstreit, and Wade Miller, who graciously agreed to author chapters to be included in this book. Your contributions are sincerely appreciated. Also, appreciation is extended to Ron Schneiderheinze, who assisted with the research effort in support of several chapters.

Finally, I acknowledge that any errors or omissions, although unintentional, are solely my responsibility.

Robert J. Birkenholz

CONTENTS

LIST OF FIGURES

LIST OF TABLES

1 HISTORICAL DEVELOPMENT

Adult education in the United States can be traced to when Squanto taught the Pilgrims at Plymouth Rock to bury a fish beside each hill of corn at planting time. Since that first lesson on crop fertilization, adult education has taken place in a variety of formats and settings. Early adult education efforts focused on teaching young colonial men to read the Bible. Subsequent adult education programs have expanded far beyond those initial efforts.

During the colonial era, settlers within a geographic region would gather periodically in town meetings. These meetings provided citizens with opportunities to discuss problems and issues confronting them and their neighbors. In addition, important news and information was disseminated to those in attendance, thus creating the first (albeit loosely organized) formal adult education activity in the United States.

Benjamin Franklin is credited with initiating several activities leading to early adult education efforts in this country. He organized the Junto around 1730 in Philadelphia as an informal discussion group of about 20 persons. The original Junto operated on a continuous basis for 30 years as a self-educating group that debated topics of morals, politics, and philosophy. Subscription libraries were created as an outgrowth of the Junto. Members pooled their resources to assemble a collection of books, which were to be read and discussed by the group. Franklin assisted in founding the Philadelphia Library Society, which was the forerunner of the public library system. It spread quickly throughout the United States. In 1833, a free town library was created in Peterborough, New Hampshire, with support provided by a municipal tax. However, it was not until the Boston Public Library opened in

1852 that the concept of a free, public library became firmly established in this country.

The Lyceum movement began in 1826 in Millbury, Massachusetts, by Josiah Holbrook. The purpose of Lyceums was to improve members through association and study, through dissemination of knowledge by the establishment of libraries and museums, and to promote tax-supported public schools. The Lyceum movement spread rapidly throughout the northeastern United States between 1826 and 1835. However, by the beginning of the 1900s the number of organized Lyceums had dwindled significantly. In 1925, it was estimated that only about a dozen Lyceums were still functioning in this country.

In 1874, the Methodist Episcopal Church organized a residential adult education activity involving a summer school program for Sunday school teachers in Chautauqua, New York. Two years later, one of the organizers, the Reverend John Heyl Vincent, published *The Chautauqua Movement*, which described how to organize and conduct a "chautauqua." The program format and purpose expanded to include the study of languages, home study courses, and a book-of-the-month club, in addition to literary and scientific circles. The original Chautauqua Institution remains in existence and is associated with Syracuse University in New York. Although early chautauquas were residential in nature, after the turn of the century, traveling chautauquas became popular by offering inspirational speakers and entertainment. Often the traveling or "circuit" chautauqua would be held in the same tent along with medicine shows, which were also quite popular during that time. By the late 1920s, the chautauqua movement had begun to suffer from the poor reputation associated with medicine shows, and the programs expired as quickly as they had grown. The Chautauqua University continues to offer correspondence courses, lectures, conferences, short courses, and other activities for adult groups from its base on the Syracuse University campus.

Following the pattern that had been established by the chautauqua movement, several higher education institutions began to develop correspondence and/or traveling educational programs for adults in their respective states. The Universities of Wisconsin, Minnesota, and California each claim to have pioneered the first program of extension and continuing education in this country. By the 1890s more than 12 state land grant universities had expanded their programs to include extension activities. In 1893, William Rainey Harper introduced adult education in a private institution by supporting extension lectures and correspondence courses through the University of Chicago. Prior to becoming the first president of the University of Chicago, Dr. Harper had been a Chautauqua Director. He later organized Joliet Junior College, another first in the history of higher education in the United States.

Federal legislation supporting adult education was first passed in 1914 in the form of the Smith-Lever Act, which created the Cooperative Extension Service based on land grant university campuses. Prior to that legislation, Dr. Seaman A. Knapp led an effort to transfer research findings from land grant colleges and agricultural experiment stations to farmers through the use of demonstration projects. Dr. Knapp was convinced that farmers would only adopt new production practices after having seen the practice demonstrated on their own or a neighbor's farm. He later became a consultant to the U. S. Department of Agriculture and helped to guide the formation of the Extension Service at the federal level, although he died in 1911 before the Smith-Lever Act was passed.

The formation of the Cooperative Extension Service resulted in the largest adult education organization ever created. The purpose of the Extension Service was to "aid in the diffusing among the people of the United States useful and practical information on subjects relating to agriculture and home economics, and to encourage the application of the same . . ." (Smith-Lever Act, 1914). During the first half of the 20th century, the Cooperative Extension Service served a critical role in linking the federal government with people in rural areas.

Programs developed to address social and economic problems in rural areas were frequently managed and delivered through the Cooperative Extension Service. Therefore, Cooperative Extension became the information conduit through which programs, such as the Agricultural Adjustment Administration, Soil Conservation Service, Farmers Home Administration, and the Rural Electrification Administration, were introduced throughout rural America.

During World War II, the Cooperative Extension Service continued to play a critical role in educating rural residents. However, since the 1960s, the focus began to shift away from farm and family management, toward leadership development, community development, socialization, and public affairs. More recently, the Cooperative Extension Service has included county staff or regional specialists who serve as resource persons to provide advice or guidance to citizens in solving local problems. Although agriculture stills holds the greatest share of subject matter specialists, other areas are becoming increasingly important, including human environmental science, 4-H and youth development, and community development.

Defining *Adult Education*

During the 20th century, the term *adult education* has been defined in a variety of ways. Knowles (1980), Newsome (1992), and Rachal (1988) noted that the multiple definitions created confusion. The term *adult education* has been used to describe the process of adult learning, an educational program designed for adults, or a collection of institutions, agencies, and organizations that provide educational opportunities for adults.

Knowles (1980) acknowledged the confusion related to defining the term in his book *The Modern Practice of Adult Education: From Pedagogy to Andragogy* where he presented three unique definitions. Knowles described the process of adult education as "all experiences of mature men and women by which they acquire new knowledge, understanding, skills, attitudes, interests, or values" (p. 25). This definition was sufficiently comprehensive to encompass planned and unplanned activities ranging in scope from formal classroom settings to reading a newspaper to catch up on news events of the day. Each of these activities is encompassed within the broad overview of adult education as a process.

The second definition proposed by Knowles assumed a more purposeful view in which adult education was defined as ". . . a set of organized activities carried on by a wide variety of institutions for the accomplishment of specific educational objectives . . ." (p. 25). The major characteristic distinguishing the latter from the previous definition is the degree of planning that preceded the acquisition of knowledge or skills by the learner. Haphazard or serendipitous knowledge attainment would not be included as part of the latter definition. Also, the contextual framework for the second definition appears to focus on a formal institution or program to deliver the information, whereas the first definition tended to include activities of knowledge transfer to mature men and women in many forms. This definition places greater emphasis on the program aspect rather than the process described in the previous definition.

The third definition offered by Knowles was more ambiguous than either of the first two, and in many ways represented a hybridization of both previous definitions. The third definition stated that adult education brought together "a discrete social system of all the individuals, institutions, and associations concerned with the education of adults and perceives them as working toward the common goals of improving the methods and materials of adult learning, extending the opportunities for adults to learn, and advancing the general level of our culture" (p. 25).

Dobbs (1970) noted that the term did not lend itself to a single definition. Rather, he indicated that adult education commonly has at least three meanings: (a) activities in which adults have and use opportunities to learn systematically under the guidance of an agency, teacher, or leader; (b) a study directed toward a critical examination of how adults acquire and use knowledge and how learning programs can be developed to assist adults to promote a mature rationality in their lives and, through these adults, in the institutions of which they are part; and (c) the learning that takes place through everyday experiences outside of organized programs. The third definition may also be characterized as random experiential learning, although some adult educators would not include this type of learning as part of adult education. Dobbs further suggested four criteria that distinguish adult learning activities. The criteria posed by Dobbs (1970, p. 2) were:

1. There is a relationship established and maintained in which the learning occurs under the direction of an educational agency, a teacher, or a leader.

2. The activities provide opportunities for learning. This means that they are planned purposefully for education and adults participate in them for that reason.

3. The learning is systematic. It results from planned educational experiences.

4. The learner is an adult.

These criteria eliminate many situations in which adults "learn" new information; however, in the absence of planned activities for the purpose of education, most definitions imply that both the learner and the leader mutually acknowledge the educational purpose of the activity.

Wagner (1993) and Riley (1993) implied that adult education primarily focused on remedial or developmental programs for persons who were not well-served by elementary and secondary schools they had attended. Adult education in that context was viewed as a second-class educational program since the content and objectives were repetitious with that taught in the public schools (Cherem, 1990).

The National Center for Education Statistics (NCES, 1994) painted a significantly different picture of adult education when it reported that in 1991 fewer than 30 percent of the workers with less than a high school education participated in adult education programs. NCES also reported that 46 percent of workers with some college and 61 percent of the workers with a college degree had participated in adult education programs to improve their job skills during 1991. These findings support other studies that have concluded that adults with higher levels of educational attainment tend to seek adult education opportunities; while adults with less education were less likely to participate in adult education. The findings reported by NCES also suggested that adult education had expanded far beyond the remedial context of early definitions.

Adult education has also been described as the fastest growing component of the educational enterprise in the United States over the past several years. Wagner (1993) described the increased attention directed toward adult literacy education and Scheneman (1993) noted that continuing professional education had expanded greatly over the past 30 years. In the spring of 1991, one out of three adults employed full-time (and one of six employed part-time) reported that they had participated in a job-related training course within the previous 12-month period (Kopka & Peng, 1994). These rates of participation, if extrapolated over the entire adult population in the United States, would suggest that of the 191 million adults in the United States, more than 61 million had participated in an adult education program or activity within the last year. This number far exceeds the 47.2 million elementary and secondary school (K-12) students (both public and private) who were enrolled in school in the United States at that time (National Center for Education Statistics, 1993).

Trends

Several factors have contributed to the increased interest in adult education. From a historical perspective, there are several trends that have become evident. Rapid advancements in technology have been a primary force contributing to the increased demand for adult education and training. This factor, coupled with a personal desire or motivation among many adults to improve their quality of life, has prompted many adults to improve their skills to qualify for higher paying jobs. Furthermore, it is apparent that many adults participate in educational activities to maintain their skills for their current position. Data from the 1991 National Household Education Survey (Korb, Chandler, & West, 1991) revealed a relationship between the level of education a person had received, the level of skill required in their job, and their participation rate in adult education. Adults who had completed higher levels of education, and/or who were employed in white-collar jobs, were more likely to have participated in adult education or training activities than persons with less education or lower-skill jobs.

This observation is cause for concern due to the increasing disparity between the education levels of American adults at the two ends of the educational spectrum. Adults with more education tend to seek out additional opportunities for learning through adult education; those with less education tend not to seek out such opportunities. The economic impact of this discrepancy will become even more evident in future years as the gap between the more educated and less educated widens.

This observation is especially troublesome in light of the intergenerational transfer effects of education and the relationship between education level and earning potential. The National Adult Literacy Study (Kirsch, Jungeblut, Jenkins, & Kolstad, 1993, p. xix) revealed that literacy was somewhat analogous to currency (in the form of human capital) in our society. Although uneducated and unskilled individuals may be able to survive in the current economic environment, the children of illiterate parents may be doomed to a state of poverty that will become increasingly difficult to overcome.

There appears to be a widening gap in the earning potential between different classes of workers in this country. Over the past 15 years, the earnings disparity between professionals and clerical workers has nearly doubled (from 47 to 86 percent). At the same time, the gap between white collar professionals and skilled trades workers has increased from 2 to nearly 37 percent (*America's Choice: High Skills or Low Wages*, National Center on Education and the Economy, 1990).

Education and training have been identified as primary factors that influence earning capacity and therefore, the standard of living enjoyed by adults and their families. The National Center for Education Statistics (1994) reported a direct relationship between educational attainment and median annual earnings, and a negative relationship with the percent of the population below the poverty line. It seems apparent that as people secure additional education, they tend to enjoy higher earning potential, which is often accompanied by a number of other quality of life benefits. Extrapolated over future generations, there looms on the horizon an ever-widening gap between the "haves" and the "have nots" in this country. Much of the difference between these two groups may be explained by the discrepancy in educational attainment. Rectifying the discrepancy between the two groups may be one of the major challenges facing adult education programs in the future. This scenario has the potential to influence the demand for adult education for many years to come.

One example of the changes expected in adult education is directly linked to efforts to modify the welfare system in this country. Since the 1930s, the federal and state governments have attempted to solve problems of poverty, unemployment, and malnutrition by providing direct financial support to individuals. History has proven the approach to be less than successful in providing a permanent solution to the targeted problems. It has become increasingly apparent (at least to some policy makers) that poverty, unemployment, and malnutrition are symptomatic of the more fundamental problem of low educational attainment. Therefore, programs are being proposed at many levels that will link welfare payments to participation in adult education. This shift in public policy will have a major impact on the public perception of the need, value, and benefits of adult education in this country. Programs created to address the educational needs of adults who are poor, unemployed, or malnourished are expected to become increasingly important in the total scope of adult education in this country.

Adult Education Programs and Sources

To grasp a more complete understanding of the scope of adult education, a graphic illustration is provided in Figure 1. Rachal (1988) provided the illustration presented in Figure 1, which uses a tree diagram. The root structure represents institutions, organizations, or agencies that provide educational opportunities for adults to acquire new information. The branches and limbs of the tree represent the major categories and types of adult education programs. This illustration may be helpful to enable the reader to examine the scope of adult

FIGURE 1.

Tree Diagram of Adult Education Programs and Sources

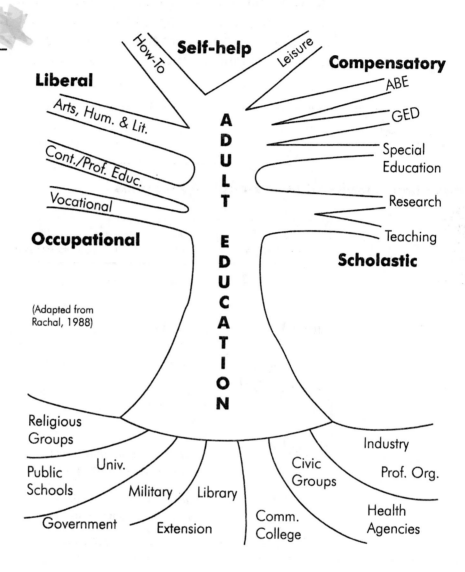

Liberal
- Arts, Hum. & Lit.
- Cont./Prof. Educ.
- Vocational

How-To   Self-help   Leisure

Compensatory
- ABE
- GED
- Special Education
- Research
- Teaching

Scholastic

Occupational

A
D
U
L
T

E
D
U
C
A
T
I
O
N

(Adapted from Rachal, 1988)

Religious Groups

Public Schools   Univ.

Military   Library

Government   Extension

Comm. College

Civic Groups

Industry

Prof. Org.

Health Agencies

education from a variety of perspectives. Even so, there may be additional options for adult learning that are not sufficiently reflected in the illustration. Rachal provided an indication of the complexity involved in defining adult education when he stated that "Adult education has such a multiplicity of forms, formats, clienteles, and purposes that attempting to taxonomize the field will continue to be a useful and inviting exercise" (p. 23).

Figure 1 reveals evidence of the multiple definitions of adult education in a graphic illustration. The programs of adult education as described by Knowles' second definition are characterized by the branches and limbs in the tree diagram. There are five major categories of adult education programs that can be differentiated by the purpose and objectives of each program. The five categories of adult education programs and the purposes of each are outlined in the following table.

TABLE 1.

Types and Purposes of Adult Education Programs

Type of Program	Purpose
Compensatory	To provide remedial learning opportunities for adults to overcome illiteracy
Liberal	To study the humanities, arts, and sciences with an emphasis on free inquiry, curiosity, and intellectual growth rather than utilitarian purposes or persuasion to partisan points of view
Occupational	To develop job-related knowledge, skills, or abilities in order to secure, maintain, or advance employment opportunities in a career
Scholastic	To teach undergraduate courses, graduate courses, or conduct research in adult learning and instruction
Self-help	To provide knowledge, information, skills, or recreational learning in order to better adjust to environments outside the work environment

Compensatory adult education programs include Adult Basic Education (ABE), General Education Development (GED), and special education programs for adults with special needs. These programs are designed to assist individuals in developing skills and abilities in order to function more fully and independently as productive members of society.

Liberal adult education programs are designed to provide self-enrichment opportunities for adults who have an interest in the humanities, arts, and literature. The ultimate purpose of these programs is to broaden one's individual thinking or perspective in traditional areas of liberal arts education.

Occupational adult education programs include courses, workshops, and conferences that are conducted to improve job related competencies of the participants. Occupational adult education programs include vocational education programs, which focus on job-related skill development, and training programs for securing, maintaining, or advancing in a career. Continuing Professional Education involves professional development programs for individuals currently employed in career fields that require professional certification or licensure.

Scholastic adult education programs encompass areas of teaching and research in which adult education is the target subject. Such program efforts are usually concentrated on college and university campuses. Participants in these programs would include individuals who have a desire to improve their knowledge and skills as adult educators.

Self-help adult education programs include a broad array of educational activities designed to improve the non-work related skills of the participants. Adult and community education course catalogs are usually filled with program opportunities addressing topics ranging from square dancing to microwave cooking and from taxidermy to insurance buying. These programs are generally of short duration and focused on a specific topic.

The root system of the illustration (Figure 1) provided by Rachal is indicative of the broad range of institutions that support or provide educational opportunities for adults. The types of institutions range from colleges and universities that tend to provide highly structured and organized programs in the scholastic category, to public libraries, which provide resource materials for use by adults who independently seek to learn more about almost any topic imaginable. Government and industry sources conduct extensive programs for employees and others on specialized topics that may be of interest to various groups and individuals within our society. Health, religious, and professional organizations often view providing educational programs as one of their primary functions. Each of the institutions listed serves a unique role in providing adult education programs; however, that role has and will continue to evolve as the needs and interests of adults in this country change over time.

Summary

The categories and types of adult education reflect the broad range of adult education opportunities that have emerged over the past several years. Individuals who have responsibility for administering, planning, funding, conducting, and/or evaluating any of these programs should develop an understanding of the principles of adult education. The remainder of this text is designed to provide a basic overview of the principles and practices of adult education for current and future adult educators.

CHAPTER 2

NEED FOR ADULT EDUCATION

The need for educational opportunities for adults in our society can be summarized in a single word "change." This word alone is the underlying factor that prompts many participants to engage in adult learning activities. *Megatrends* (Naisbitt, 1982) was among the first of many publications that described how American society was being transformed from its historical foundation based on constancy and tradition, toward a new paradigm based on dynamics and constant change. Naisbitt suggested that in the new information society, we must learn from the future in the same ways that we have learned from the past. We must learn to anticipate. He went on to identify five factors that should be kept in mind while contemplating the shift from an agricultural to an industrial and more recently to an information society.

1. The information society is an economic reality, not an intellectual abstraction.

2. Innovations in communications and computer technology will accelerate the pace of change by collapsing the *information float* (i.e., the amount of time information spends in the communication channel).

3. New information technologies will at first be applied to old industrial tasks, then, gradually, give birth to new activities, processes, and products.

4. In this literacy sensitive society, when we need basic reading and writing skills more than ever before, our education system is turning out an increasingly inferior product.

5. The technology of the new information age is not absolute. It will succeed or fail according to the principle of high tech/high touch. (Naisbitt, 1982, p. 11)

Each of the propositions outlined above has and will continue to have important implications for adult education. The absolute value of information is greater than at any previous time in history. However, simply having access to information is not sufficient. We must have information in a timely manner, in a usable format, and in concise terms to facilitate application when and where it is needed. Analogous to just-in-time inventory management practices, which have become popular in the manufacturing sector; adult learners will expend their resources of time and money when they "feel the need" to learn.

During the agricultural age, wealth was a function of the size of the farm or ranch that was owned (or at least controlled) by an individual. The industrial age witnessed a shift to where control of capital assets was the primary feature that determined the wealth of an individual or firm. We are shifting to an environment in which information is the measure of wealth. However, possession of or access to information is only part of its value. In addition, timeliness and communication ability also influences the value of information. We are entering an era in which the sheer quantity of data and information at our fingertips is overwhelming. However, the ability to translate the volumes of data into meaningful information, and ultimately, usable knowledge, is an important value-adding capability for adults to recognize and develop. Furthermore, adult educators need to provide educational programs designed to equip and empower adult learners with skills of data analysis, synthesis, and discernment.

The systems of communications constructed in recent years are truly remarkable. Most recently, the increased accessibility of information databases on the Internet system has opened vast resources of information to anyone with the capability and interest in "surfing the net." Not only has the accessibility of information changed; but the speed with which information can be transmitted from one point to another has increased at an exponential rate. Communication technology changed more in the past 25 years than in all previous recorded history. Innovations, including FAX machines, computer modems, satellite and microwave transmission, fiber optic cable, FM-band radio, video tape, CD-ROM, and plain paper copiers, are a few examples of communication technologies introduced on a broad scale in recent years.

Another aspect of the information avalanche that surrounds us in the modern age is the speed with which we are expected to make decisions. There has been increased emphasis in recent years on developing skills of decision making and problem solving. These skills involve the analysis and application of information to specific problem situations. However, the ability to sift through information, retaining that which is useful and discarding that which is extraneous, has become an important skill in this information age. In previous generations, the problem-solving dilemma was a lack of information needed to solve a problem. Today, the problem is less often a lack of information, and more frequently is a problem of discernment; i.e., what information is applicable and what is not?

In addition to the exponential increase in the quantity and accessibility of information, there is an increasing problem that may be characterized as *information aging*. Libraries and other information depositories are becoming filled with volumes of information that, at one time, was current, accurate, and relevant. However, in the advancing information age, we find that frequently the information may have *aged* to the point it is no longer factual or relevant. Therefore, we are entering an era in which concepts and theories may have been viewed as accepted norms at one point, but no longer hold true in today's world.

The concept of information aging is also a concern for an individual's knowledge base. What may have been learned previously, may later be found to have changed, or even proven to be untrue. Therefore, not only do adults need to concern themselves with learning new information in an attempt to maintain a level of currency in a given field; but they must also develop the ability to "unlearn" information that may no longer be accurate or appropriate. In the past, one frequently heard adults say "You can't teach an old dog new tricks." However, in today's society, not only do adults need to learn new tricks; but they also need to unlearn many of the old habits, beliefs, and facts that are no longer valid. The obsolescence of knowledge is becoming as much a fact of life in the information age as the need for lifelong learning.

Changes in the job market are another factor that has contributed to the need for adult education. The demographics of U.S. society reveals the average age in this country to be increasing. In addition, there are more adults entering or re-entering the work force and fewer young people seeking jobs. Therefore, many of the new jobs that will become available or be created in the next few years will be filled by adults who are already in the work force. Many adults who will assume the new jobs will also need to develop new or upgrade their current skills to be successful (Darkenwald and Merriam, 1982). For example, nurses, lawyers, or teachers who may have discontinued their careers in the 1970s, and attempt to re-enter those fields in the current environment, would need extensive retraining to be competitive and successful. The training and retraining of adults to maintain and/or advance in their jobs will continue to place a high demand on adult education programs in the future.

The demand for adult education presents an interesting paradox when examined on an individual basis. Follow-up studies of adult education participants have revealed a direct, positive relationship between the level of education they have attained and the likelihood of their participation in adult education activities. Adults with more education have a stronger tendency to participate in adult education activities than those who have less education. This situation may be illustrated in two-dimensional space using the image of a sphere (see Figure 2). The inside of the sphere represents the knowledge that an individual possesses. The area outside the sphere represents the sum total of all

FIGURE 2.

**Individual
Knowledge**

Universe of Knowledge

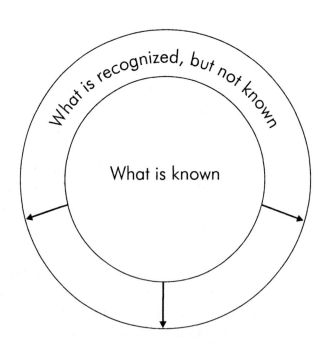

knowledge in existence at any time, but unknown to the individual. The interface, represented by the surface area of the sphere, depicts the knowledge that an individual recognizes the existence of, but does not know.

The paradox occurs as an individual acquires more knowledge (i.e., enlarges the area inside the sphere). Mathematically, the surface area of the sphere increases at an exponential rate relative to the increase in the amount of knowledge added (inside the sphere).

This relationship may be helpful in visualizing the reality of the situation in which individuals with more education tend to seek out additional educational opportunities while those with less education are less inclined to do so. As people expand their knowledge base, they also increase their awareness of what they do not know. Although education expands the base of knowledge, it also tends to expand people's sense of ignorance or inadequacy (i.e., recognizing there is even more to the universe of knowledge that they do not know).

Adults with a more limited knowledge base do not realize there is a vast quantity of information that is unknown to them. They may not recognize (or appreciate) the fact that there are unexplored worlds of information that are just beyond the boundary of their consciousness.

This concept of individual knowledge acquisition somewhat parallels the scientific research process. Frequently, researchers enter into research activities to find answers to previously unanswered questions. However, quite often, the outcome of such investigations results in even more unanswered questions than were originally posed. However, the new questions tend to be better (i.e., more informed or precise questions) than those initially targeted in the research project.

Returning to Figure 2, the surface area of the sphere represents the number of unanswered questions that confront an individual. Persons with more extensive knowledge bases tend to have more unanswered questions to motivate them to seek answers (i.e., participate in educational processes). However, obtaining answers to those questions tends to further expand people's knowledge base, which in turn results in them asking even more questions; and the process continues in an ever-widening spiral of adult learning.

Unfortunately, the converse situation also occurs among adults with limited knowledge bases. These adults tend to recognize fewer unanswered questions that prompt them for solutions, and therefore, their educational spiral is much more restricted. This situation is one of the major challenges that must be overcome in addressing the problem of adult literacy in this country. Many times, adults with inadequate reading, writing, and computational skills fail to recognize how such knowledge and skills may be utilized to improve their lives. Lack of basic skills is one of the contributing factors to the downward educational and economic spiral, which faces the majority of Americans who are handcuffed by illiteracy. If the chain of illiteracy could be broken at some point, the opportunity for adults to spiral upward in educational and economic terms becomes much more feasible (National Center for Education Statistics, 1994). The latter scenario is becoming one of the guiding principles regarding the reformation of welfare subsidy programs in this country. Recent policy and program shifts have been directed toward encouraging adults to participate in adult education programs as a prerequisite to receiving welfare benefits. This policy shift is being implemented to change the welfare system from a "way of life" to a temporary safety net for the truly needy. As such, adult education is being viewed as an important vehicle to transport individuals and families from poverty levels to higher and more acceptable standards of living.

The need for adult education in our society should also be examined in the context of the demand for adult education. In the previous chapter, five types of adult learning opportunities were introduced. Each of the types were distinguished from one another with respect to the function to be served by each. Occupational programs were designed to enhance career skills for participants. Self-help programs were conducted to address the recreational and leisure time desires of adults. However, each of these program offerings should consider the needs of adult participants in planning, organizing, conducting, and evaluating the activity.

Determining the needs of adults will be described in greater detail in Chapter 7; however, at this point, it is important to recognize that adult education should be designed to address the needs of the participants. Human Resource Development (HRD) programs in major corporations develop training programs for their employees based on the needs of participants; however, the needs may be determined by supervisors, managers, or other administrative personnel rather than directly by the participants themselves. In such a case, the needs are identified by a third party.

Participant needs may also be determined by an adult educator as in the case of a program designed to teach adults how to play contract bridge. In this case, adult participants may lack the knowledge needed to accurately assess what they need to learn in order to play such a complex card game. Therefore, the adult educator is empowered with the authority to define the educational needs of the participants.

A third type of educational need to consider would be that of the self-directed learner. Self-directed adults tend to be intrinsically motivated, define their own learning objectives, and take the steps necessary to achieve them. Completely self-directed learners may assume the role of teacher and learner simultaneously. In so doing, they have the inherent responsibility to plan, organize, conduct, and evaluate their learning activities. Self-directed learners are becoming more commonplace in our society. One trend that will have a dramatic affect on adult education programs in the future is the extent to which individuals embrace the concept of self-directed and life-long learning.

CHAPTER 3

CHARACTERISTICS OF ADULT LEARNERS

Adult learners possess a number of characteristics that distinguish them as a group from younger learners. Zahn (1967) illuminated this point by noting that adults were not just large children in that they differed from youth in their learning in several ways. Although this is a generally accepted premise among adult educators, most would also agree that there are also a number of characteristics that adult and adolescent learners have in common. This chapter provides an overview of the characteristics that adult learners have in common with their adolescent counterparts, and those characteristics that are unique to adult learners.

Thorndike (1929) has been credited as the first researcher to provide empirical evidence related to the learning ability of adults. In his initial studies, Thorndike reported that adult learning ability was similar to that of a late teen. He further suggested that adult learning ability was nearly constant throughout middle age (50 years) and decreased thereafter at a rate of approximately one percent per year due to a decrease in sensory acuity (i.e., sight, hearing, taste, smell, etc.). Subsequent studies by Miles and Miles (1932) and by Jones and Conrad (1933) extended the constant learning curve to well beyond age seventy. Therefore, one of the most significant principles to consider in adult education is the belief that adults maintain their inherent ability to learn throughout life. Adult educators, to be successful, need to fully embrace this principle of adult learning as a foundation of their overall philosophy of adult education. Believing that all (or at least the great majority) of adults possess the ability to learn is a significant premise from which to begin to examine the field of adult education (Bergevin, 1967).

Prior to examining the characteristics of adult learners in any detail, it is necessary to describe the context of what is meant by the term *learning*. How do people learn? How do we know when someone has learned something? What should adult educators do to facilitate the learning process for adults? These are important questions that need to be examined by adult educators.

Learning has been defined by educational psychologists as a *change in behavior*. Most educators further restrict the definition to include only those behavioral changes that are the result of personal experience, rather than behavioral changes brought about by natural maturation, illness, fatigue, or drug inducement.

Learning, as a process, may involve both planned and unplanned activities. In either case, the change in behavior (whether positive or negative) may be classified as learning. One adult issued a ticket for exceeding the speed limit on the highway "learns his/her lesson" and avoids driving too fast in the future. However, another person who receives a ticket may continue to drive too fast, apparently not learning from the experience. The differences in the outcome (i.e., learning) that occur when two individuals have the same experience are difficult to explain. However, it is important to note that learning involves significantly more than a simple input:output relationship.

Adult Learning Theories

Numerous learning theories have been promoted in the educational literature. Although many of the theories have been described in the context of adolescent and youth education, Merriam and Cafferella (1991) outlined four major learning theories that have applications to the field of adult education.

The *Behaviorist Orientation* is founded on the belief that learning is a direct result of the connection between a stimulus and a response. Thorndike outlined three laws of learning that support the behaviorist learning theory. The "Law of Effect" stated that learners repeat certain behaviors (i.e., responses) to experiences that produce pleasing results.

The "Law of Exercise" stipulated that a change in behavior results when an individual observes the connection between a stimulus and a response on a repetitive basis. The "Law of Readiness" noted that learning is enhanced if the adult learner is mentally prepared to recognize the connection between the stimulus and the response. B. F. Skinner, a well-known behavioral psychologist, proposed to expand the behaviorist theory to encompass operant conditioning in which positive reinforcement was recommended for desirable behavior, and negative behavior was to be ignored. Skinner hypothesized that behavior that was continually ignored would eventually be discontinued.

Another major premise that undergirds the behaviorist theory is that all behavior is learned and, therefore, all behavior can be modified or changed by further learning. This view is a powerful force in adult education, although there are other learning theories that are based on a different set of beliefs.

The *Cognitive Orientation* was originally proposed by a Gestalt psychologist (Bode, 1929). The cognitive theory suggests that learning is a result of individually centered mental functions. Information processing involves the steps of acquisition, transformation, and application of new information in a new context. The distinguishing feature of cognitive learning theory is the internal locus of control. Cognitive theorists believe that individuals control their learning through their own mental processes, whereas behaviorists view the locus of control over the learning process coming from the individual's environment.

The *Humanist Orientation* to learning is primarily based on the research and writings of Abraham Maslow and Carl Rogers. Maslow is considered to be the founder of humanistic psychology and is noted for the development of the Hierarchy of Human Needs (see Figure 3) to explain the motivations behind adult behavior. Humanist theory was founded on the belief that learning (i.e., behavior) is directed by an intrinsic desire or motivation to achieve one's fullest potential. While behaviorists view control coming from the environment, cognitivists view control coming from the individual's own thought processes, and humanists view the control mechanism to originate within the individual's subconscious. Maslow's Hierarchy of Human Needs is the basis for motivation theories prevalent in adult education from the humanistic perspective. Each person is assumed to have a subconscious desire to fulfill each level of prescribed needs in a hierarchical fashion.

Lower level needs that have reached an acceptable level of fulfillment cause individuals to begin seeking fulfillment of the next higher level of needs. This orientation to adult learning theory may be especially applicable during adult transitions, such as job changes, marriage, divorce, child birth, etc. Major and unanticipated changes in adult life frequently cause individuals to drop back to where they must take steps to ensure fulfillment needs at lower levels of the hierarchy, which, prior to the life-changing event, had already been fulfilled.

FIGURE 3.

**Maslow's
Hierarchy of
Human Needs**

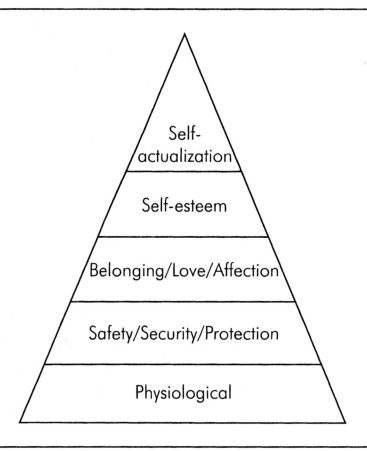

In the context of the humanistic orientation, adult educators need to assess individual participants to determine the lowest level on the hierarchy at which each persons' needs are unfulfilled. Recognition of individual needs should provide a signal to an adult educator with regard to an appropriate course of action to facilitate the learning process. Examples of alternative responses or activities that may be appropriate for individuals at each level of Maslow's Hierarchy are outlined in the following table.

TABLE 2.

Adult Education Opportunities to Fulfill Human Needs

Level of Maslow's Hierarchy of Human Needs	Opportunities for Fulfilling Needs
Self-actualization	Art, music, humanities, bible study, volunteering (scouts, hospital, nursing homes, 4-H, day care, etc.) Habitat for Humanity
Self-esteem	Assertiveness training Toastmasters, public speaking Dale Carnegie course Personal improvement seminars Parenting classes
Belonging/Love/Affection	Leisure activities Group/organization membership Team building Cooperation Leadership training
Safety/Security/Protection	Job skill training/re-training Fire safety Buying a home Selecting a life insurance program Self-defense training
Physiological	Free health-care screening Soup kitchens Food stamp programs WIC programs Pre-natal health checkups

According to the humanistic orientation, people are motivated to learn due to an intrinsic desire for self-improvement. Upon fulfillment of each level of need, the theory suggests that a subconscious drive will encourage them to strive for fulfillment of higher level needs.

The fourth major orientation to adult learning is the *Social Learning Theory*. This orientation suggests that adults learn through observation, and upon reflection, will imitate or modify their own behavior accordingly. Bandura (1986) suggested that adult learning is a direct result of the interaction between adults and their environment. The social learning theory implies that the context or environment plays a significant role in adult learning and may have more explanatory power than the three adult learning theories previously presented. The social learning theory appears to provide insight into the concept of acquisition of adult roles, leadership/management training, and mentoring activities for adults.

Differentiating Adults from Adolescents

Adult learners can be differentiated from adolescents in at least four ways. Adults have responsibilities associated with (a) being good citizens (b) maintaining economic stability for themselves and their families; (c) serving in the roles of parents and adult family members; and (d) having the responsibility to transmit social, cultural, and spiritual values to future generations. In addition to situational and demographic characteristics that distinguish adult learners from adolescents, adults generally participate in learning activities on a voluntary basis, whereas adolescents frequently face mandatory attendance policies. Having the option to "vote with their feet" (Imel, 1994) by choosing whether or not to participate in educational programs significantly alters the frame of reference for those in attendance. The environment created when attendance is required by some authority other than the actual participant often results in a less-than-ideal learning situation. Adult learners who choose to participate in educational programs frequently bring with them a high level of motivation that reduces the need for the adult educator to employ motivational techniques to stimulate the participants to want to learn. Rather, the challenge for the adult educator becomes one of assessing and monitoring adult participants in order to help direct their energies in appropriate channels to achieve the desired outcome. Having a motivated learner is a distinct advantage that is recognized by experienced adult educators. However, there are some adult education programs that are not voluntary. In these situations, adult educators face some of the same challenges as secondary teachers in attempting to motivate students to learn. Therefore, voluntary participation in most adult education activities is a distinct advantage that contributes significantly to successful adult learning.

Adult Basic Education (ABE) and General Education Development (GED) programs for prison inmates may be attended by persons whose motivation ranges from that of a totally self-directed learner to individuals whose involvement is a prerequisite to their eventual parole. Therefore, adult education programs may involve participants with widely divergent motives. Some participants may exhibit a high level of intrinsic motivation, whereas others may have ulterior motives that prompt their attendance but do not increase their desire to learn. Recognizing the factors that influence participant motivation for adult involvement in educational programs is an important prerequisite to facilitating the learning process with adults.

Adult Learning Ability

Several studies undertaken in the 1900s examined the ability of adults to learn. Thorndike (1929) is frequently cited for conducting the seminal study that concluded that human learning ability increased dramatically throughout adolescence and young adulthood, reaches a peak at about age 25, and then declines at a rate of less that one percent per year to age 50. Miles and Miles (1932) and Jones and Conrad (1933) extended the age of decline to 70 years and beyond. Although the decline in learning ability had been observed, later studies more accurately partitioned the "rate" of learning from the "ability" to learn. Subsequent studies began to control for decreased reaction speed, eye sight, hearing, and other sensory performance characteristics and concluded that there was virtually no decrease in learning ability, only a decrease in the rate of learning due to decreased sensory acuity. Furthermore, after accounting for increased levels of intrinsic motivation, greater attention span, and a more extensive experience base, adults were found to have superior learning ability to adolescents in several areas.

Early studies in adult learning attempted to predict changes in learning ability as a function of chronological age. However, due to difficulties associated with planning and conducting longitudinal studies with intact cohort groups, early researchers had employed cross-sectional designs to collect data, which were representative of adults from specific age groups in the population (e.g., age 20 to 29, 30 to 39, 40 to 49, etc.). The data collected in these studies clearly revealed that adult learning appeared to decrease as age (the independent variable) increased. However, the design did not control for differences in the amount of formal schooling that had been completed by each age

FIGURE 4.

**Adult Learning
Ability**

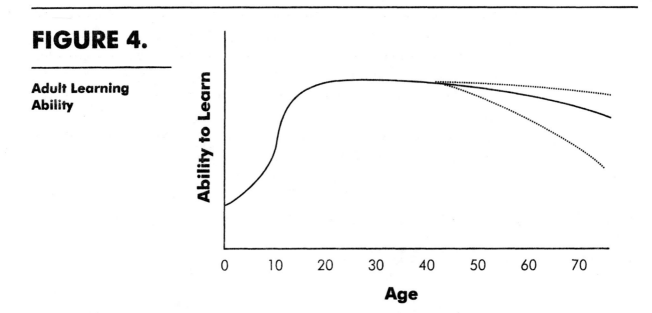

group. Subsequent research has revealed that years of formal education completed is related to learning ability. Therefore, the decline in learning ability, which was previously thought to have been a function of the aging process, may have been masked by an intervening variable related to the years of formal schooling completed.

Longitudinal studies have since been conducted that provide evidence suggesting that learning ability does not vary significantly after reaching its peak at about age 25. Rather, the ability to learn remains relatively constant throughout an individual's adult life, and changes in learning ability tend to be a result of major life transitions related to changes in situational characteristics, personal values, goals, and individual self-concept.

In addition to the stabilization of learning ability of persons throughout their adult life, there is also a stabilization of attitudes, values, and interests. Adolescents and young adults are frequently characterized as searching souls whose attitudes, values, beliefs, and interests vary considerably over a relatively short period of time. However, it has been found that as individuals age, educational activities and experiences tend to crystallize their perception of their abilities and the environment. Adults with a positive self-concept will tend to maintain that perspective throughout their adult life. On the other hand, adults who have more negative views of themselves and their abilities and of the world in which they live will likely become even more pessimistic as they grow older. It has been suggested that aging alone may not bring about more conservative views, and therefore, greater resistance to change; but rather it may reflect increased stabilization of values, interests, goals, and beliefs. This observation when viewed through the eyes of a younger generation may be interpreted as increased conservatism associated with the process of aging. Alternatively, the reluctance to change may be the result of individuals who, as they get older, have more clearly defined their values and goals than the members of the younger generation.

Summary

Adult learning ability appears to be fairly constant from adolescence through middle age; however, the rate of learning may decline due to sensory deterioration. Adult learning tends to adhere to the self-fulfilling prophesy that "If you think you can, you can, and if you think you can't, you're right!" Most adults have the ability to learn whatever they set their minds to learn. Conversely, without the proper motivation, no amount of effort can force anyone to learn.

CHAPTER

ADULT
...ARNING PRINCIPLES

...er presents principles that support the practices used in ...ducting, and evaluating adult education activities. To ...adept at facilitating adult learning, adult educators must ...nd principles related to the learning process. The adult ...ciples described in the following paragraphs provide ...decision-making relative to adult education programs; ...e is one caveat. Each of the principles described can be ...has been refuted by some adult educators in some con-...universal truth regarding principles of adult learning is ...no universal truths. Therefore, these principles should ...as hard and fast laws governing adult learning and adult ...ograms. Rather, these principles should be thoughtfully ...h reference to each specific context or application. Those ...have merit should be further considered, and those that ...ould be temporarily set aside. The value of this set of prin-...the questions and considerations that may be prompted ...h unique situation. Adult educators need to consider each ...ative to a specific environmental context, to determine its

Principle 1   LEARNING IS CHANGE

Learning has been defined in a number of ways, but most frequently it is functionally defined as a *change in behavior*. Often, we anticipate that the behavioral change resulting from learning (especially with adolescents) will be observable and measurable. This assumption may not be fully evident for all individuals, especially in the short run. Often, learning situations will not produce externally observable changes in behavior, but may have an imperceptible effect on a person's attitude, disposition, or predisposition. Changes in a person's thought processes or attitude toward a topic may not be fully recognizable (even by the individual learner) in the short term; but when examined at a later time, a change in behavior may be traced back to a prior learning experience.

Change in knowledge is probably the most common result of adult learning activities. Adults are continuously confronted with learning situations as they interact with other persons, read newspapers and magazines, and generally participate in the activities of adult life. Adult learning, from the viewpoint of acquiring new information and knowledge, is a part of everyday life. However, some individuals are proactive in their quest to seek out learning opportunities, while others are quite content to avoid internalizing the mass of new information that bombards each of us on a daily basis. From this perspective, *learning is an individual process*. Although groups of individuals may participate in a learning activity, the net result (i.e., what is learned) will vary widely among the range of participants.

Principle 2   ADULTS MUST WANT TO LEARN

Adults usually have the freedom and maturity to choose whether or not to become involved in learning activities. However, in some instances, adult education programs may be conducted where participants are encouraged or even required to attend. The potential for learning diminishes markedly for adults who are required to participate by some authority figure. Learning efficiency and achievement is directly correlated with the individual's motivation for participation. Persons who are highly motivated and voluntarily elect to participate in educational activities are usually much more successful and efficient in learning new information or skills. Individuals who lack internal motivation are often reluctant to participate in learning activities, which presents a barrier to learning.

Participants who become involved in adult education activities in response to a requirement (i.e., inservice, certification, update training, etc.) may have little or no desire to learn the information being presented. Often, adults participate in learning activities for reasons that differ widely from the intended purpose and objectives of the program. Therefore, adult education program planners must be prepared to identify or anticipate factors associated with adult participation in adult education programs. After identifying factors contributing to adult participation in learning activities, adult educators can plan programs that will more closely address the needs and interests of participants.

Programs involving adult participants who are not intrinsically motivated, should include activities that enable participants to develop a "felt need" for the information being presented. Addressing the rhetorical question of "Why are we here?" is very important in establishing a common bond linking program planners and participants. Only after participants concur that there is a "need" for them to learn, will learning occur. The adage "You can lead a horse to water, but you can't make it drink" is applicable to teaching adults who are not motivated to learn.

Principle 3   ADULTS LEARN BY DOING

Adults learn best through direct participation in the learning process. Although this is a universal truth for learning in general, there is one caveat. Many adults will not take the initiative to fully engage in learning experiences without encouragement. Therefore, adult educators must be skilled at the art of facilitating participation by planning programs that begin at the "entry level" of participants, but increasingly provide opportunities for the adults to become engaged in the learning process.

Overcoming inhibitions that prevent adults from engaging in learning activities is important to the success of adult education programs. Many adults have low self-esteem, lack confidence in their abilities, and fear ridicule or failure. These characteristics are significant barriers to adult learning that should be recognized and overcome in order to encourage participation. Adult educators should organize activities during early stages of the program that will virtually guarantee success in order to overcome such psychological barriers to learning. Once adult participants feel comfortable with their active participation, they will become more liberated from the inhibitions that interfere with their learning, and can successfully engage in more challenging learning activities.

Principle 4   ***LEARNING SHOULD FOCUS ON REALISTIC PROBLEMS***

Adult learning should focus on knowledge and skills that have immediate application. The sooner new knowledge and skills are put into practice, the better. Most adults do not have the time, patience, or inclination to learn information that lacks relevance. Therefore, **adult educators should address topics and issues of relevance in the very beginning.** This eliminates the participants asking (even rhetorically) "What does this information have to do with me?"

One strategy frequently employed in adult education has been termed the *inductive approach* to teaching (see Figure 5). With this strategy, adult educators use realistic situations (e.g., case studies, role play situations, demonstrations, etc.) to engage participants in identifying and defining the problem and proposing alternative solutions. After two or three situations have been examined, the adult educator guides the participants through a discussion in which they reach a conclusion that involves a general principle or concept. This strategy tends to be more effective with adults than using the deductive approach, which

FIGURE 5.

Inductive and Deductive Approaches to Teaching

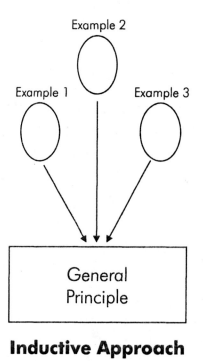

Inductive Approach

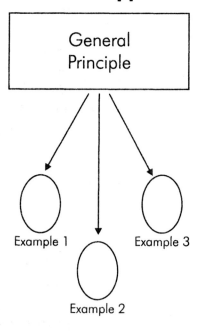

Deductive Approach

characterizes the teaching strategies employed by many secondary educators.

The *deductive approach* involves presenting a general principle, concept, or hypothesis followed by a presentation or discussion of multiple applications. In the inductive approach, learners are led from a specific example to a general concept. With the deductive approach, the reverse is true; learners are led from a presentation of a general concept toward specific applications. The two instructional strategies differ with regard to assumptions about the learners. Adult educators should recognize that the level of prior experience among the participants will influence the utility and effectiveness of each strategy.

Participants with prior knowledge and experience will usually respond favorably to the inductive approach. However, use of the inductive approach with an audience that lacks the necessary background and experience to relate in a meaningful way to the examples presented may result in a "pooled ignorance" outcome. Participants with little or no experience with the target subject matter would benefit from the use of the deductive approach in the teaching/learning process.

The role of the adult educator differs greatly depending upon the teaching strategy employed. Using the inductive approach, the adult educator is more likely to employ questioning strategies to encourage participants to arrive at a conclusion based on a discussion of multiple scenarios. The deductive approach requires that the adult educator be more adept in explaining the principle, concept, or hypothesis to participants, followed by applications to specific situations. Again, the level of participant experience should be a primary factor influencing the selection of the appropriate teaching strategy.

Principle 5   *EXPERIENCE AFFECTS ADULT LEARNING*

Experience is an important factor that has a tremendous influence on adult learning (Dewey, 1938; Lindeman, 1926; & Kolb, 1984). The effect can be either positive or negative, or possibly both in the same setting. Experience is a cumulative characteristic, although the amount and quality of experience can vary widely among individuals. Participants enter into adult education activities with their individual "set" of previous experiences. These can be viewed as a partially constructed

FIGURE 6.

**Foundation for
Learning**

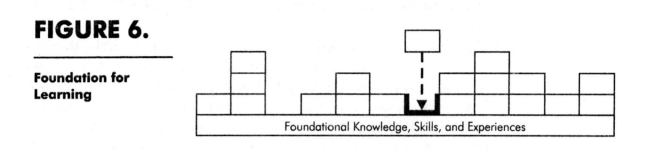

Foundational Knowledge, Skills, and Experiences

brick wall (see Figure 6). Some individuals have a broad and firm foundation underlying their wall of knowledge, skills, and experience. Others may be lacking in the scope and quality of foundational experiences. As individuals acquire new information and skills, they do so in the context of their past experience. Adding a new "brick" of knowledge or skill to a participant's wall is influenced by life experiences as well as by prior experience with learning. Therefore, adult educators should recognize how experience (whether positive or negative) affects adult learning. Positive experiences will frequently enhance the quality and efficiency of learning, whereas negative experiences have an inhibiting effect on learning. In the latter situation, adult educators may need to work cautiously to remove or disassemble portions of an individual's "wall of knowledge and experience" (i.e., unlearning) to allow the new knowledge or skill to be placed in the appropriate context.

The process of unlearning is becoming more important in today's society. What was once viewed as "truth" in days gone by, in many cases no longer holds true in the current environment. Adults who recognize and have the ability to unlearn and relearn will be more successful in the fast-changing environments of the future. Adult educators should be prepared to facilitate the adult learning process and unlearning process simultaneously.

Principle 6   *ADULTS LEARN BEST IN INFORMAL ENVIRONMENTS*

One of the major barriers that inhibits adult learning is the memory of rigid adherence to rules during formal schooling as a child. Overcoming such mental barriers is an important prerequisite to full empowerment of adults to become fully immersed in the educational process. Most adults are accepting of behavioral guidelines, even more so if they have input into the development of such guidelines or poli-cies. Providing opportunities for adults to establish their own set of operating rules and policies not only empowers them as self-directed learners, but it also enables them to focus on positive aspects of what participants are expected to do, rather than the negative approach of reflecting on what participants should not do. The wording and tone of guideline and policy statements should be constructed carefully since they have an important psychological effect on how adults view the educational climate. Rigid and strict rules remind many adults of their school days as a youth, whereas broadly stated guidelines project more of a sense of flexibility within limits, which tends to be more conducive to adult learning. As a rule, it is best to keep the number and scope of rules to a minimum.

Refreshments are one significant demarcation between adult educa-tion and the schooling of adolescents. Scheduled refreshment breaks in adult education programs can be used to enhance the learning process by providing opportunities for participant interaction between and among themselves, the adult educator, and resource persons who may be in attendance. Refreshment breaks also provide opportunities for adults to ask clarifying questions, which they may have refrained from asking in a larger group or more formal setting. These opportunities are frequently revealed as the times when true learning occurs. There-fore, adult educators should recognize the learning opportunities that exist during informal interaction and strategically plan for such oppor-tunities in their programs.

Principle 7   USE VARIETY IN TEACHING ADULTS

Like children, adults learn through their senses. Research has shown the more senses involved in the learning process, the more effec-tive the learning experience. Therefore, when planning adult education programs, it is wise to incorporate activities that require adult partici-pants to utilize the maximum number of senses (i.e., seeing, hearing, feeling, tasting, etc.). Involving multiple senses in the learning process produces an increase in the information retention rate in the short run, but an exponential increase in long-term retention. Therefore, using a variety of instructional methods and strategies will greatly increase the effectiveness of the instructional effort. However, there is usually a trade-off between instructional efficiency and instructional effective-ness in education. Methods that maximize effectiveness (i.e., long-term retention) tend to be less efficient, and methods that are more efficient (e.g., transmit large quantities of information in a relatively short time) tend to be less effective. Adult educators should recognize this trade-off and make instructional planning decisions with regard to the overall purpose and objectives of the educational activity. Generally, adult educators are forced to choose some middle ground, compromising between a desire to maximize educational effectiveness while responding to the need to be efficient in the process.

Principle 8   ADULTS WANT GUIDANCE, NOT GRADES

Adults are curiously individualistic when it comes to evaluating their achievements or performance. Most adults do not enjoy being held up as an example to others out of fear of humiliation or ridicule. However, they generally have a desire for some external affirmation of their progress relative to their peers.

Adult educators should avoid the application of rigid, external performance standards except in areas that have prescribed certification or licensure requirements. Most adults are accepting of suggestions for improvement regarding their performance, provided it is offered in a tactful and non-threatening manner. However, self-evaluation among adult education participants is the most relevant measure of achievement in terms of their individual desired outcomes. Adults participate in educational activities for different reasons, and their individual standards for self-evaluation will differ accordingly.

Adults who have been absent from any formal learning environment for a number of years will often benefit from signals of encouragement and affirmation of their capabilities as learners. New adult learners frequently struggle with overcoming intense feelings of self-doubt. Therefore, adult educators should be constantly aware of cues provided by adults who lack confidence in their abilities as learners, and provide support and encouragement to help them progress towards their goals. Failure to recognize signs of discouragement will often result in adult dropouts, which may inhibit the adult participant from attempting formal learning again in the future. The ability to diagnose and prescribe solutions for adults who lack confidence in their ability to learn is an important skill for adult educators to develop. However, the first step in developing that skill is to recognize the "need" for the knowledge and skill in the first place.

These eight principles provide a basic overview of adult education guidelines to consider in planning and conducting programs for adult learners. These principles may be useful in the formation of a general statement of philosophy regarding adult education (see Appendix A). Developing a written philosophical statement consisting of beliefs, attitudes, and values can be very insightful as adult educators reflect on their role in facilitating the adult learning process. Zemke and Zemke (1981) also developed a list of "30 Things We Know for Sure About Adult Learning" (see Appendix B) that presents another perspective of adult education principles that would be useful to consider in planning educational programs for adults. Appendix C (Principles of Teaching and Learning) and Appendix D (Variables of Effective Teaching) should also be reviewed as quidelines related to teaching and learning with adults. Although these materials were designed for elementary and secondary educators, there are implications for adult education as well.

Andragogy and Pedagogy

Knowles (1980) has been credited with introducing the term *andragogy* into the adult education vernacular in the United States. He defined the term as the "art and science of helping adults learn" (Knowles, 1980, p. 43). This definition was originally developed to contrast the term *pedagogy*, which was defined as the "art and science of helping children learn." Although the terms *pedagogy* and *andragogy* were originally cast as dichotomous approaches to teaching and learning (see Appendix E), more recently the terms have been judged to represent a "continuum" ranging from teacher-directed to learner-directed approaches, respectfully (Merriam, 1993).

Selecting the best approach to employ in planning and conducting educational programs for adults has little to do with the chronological age of the participants. A major consideration in planning adult learning activities is the maturity of the learner. "Maturity" in this context does not refer to the learners' chronological age or their maturity as an adult in society. Rather, in this context, maturity is a function of the adult participants' ability to self-direct their own learning. Other factors affecting learner maturity include the amount of knowledge in the area already possessed by the learner, the level of interest in and the need to acquire the knowledge or skill, the willingness of the learner to accept responsibility for their learning, and the level of learning skills

FIGURE 7.

Four Stages of Adult Learning

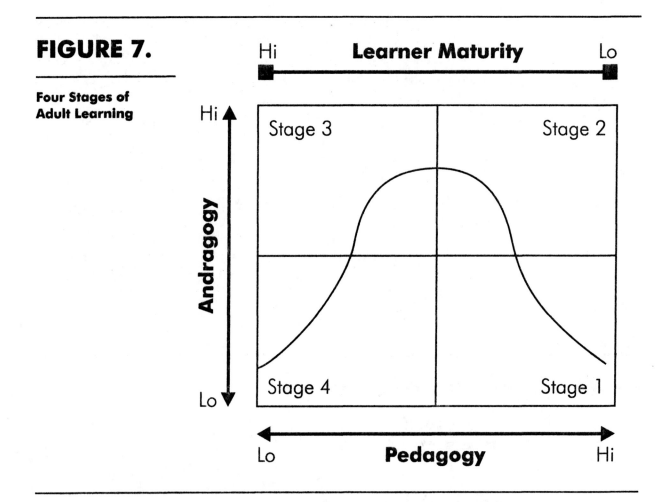

and experience possessed by the learner (Smith & Delahaye, 1987). The concept of learner maturity is illustrated by the two-dimensional graph in Figure 7 (see Stuart & Holmes, 1982; Delahaye, Limerick, & Hearn, 1994; and Grow, 1991).

Stage 1   *HIGH PEDAGOGY, LOW ANDRAGOGY*

The first stage of learner maturity is characterized by learning environments that are highly structured (i.e., pedagogically speaking), with little or no opportunity for input from the learners (i.e., low andragogy). Examples of such learning environments in adult education would include licensure or professional certification (or re-certification) programs in which the performance objectives or outcomes are predetermined and imposed by a third party agency or institution. Participation in such programs may be a requirement for continued employment in the career field. Learners in this context may participate in the learning environment, but may display a lack of intrinsic motivation in the process. Adult educators in these situations should develop and implement strategies to enhance the active involvement of participants in the learning activities.

Stage 2   *HIGH PEDAGOGY, HIGH ANDRAGOGY*

Programs that are designed for learners in the second stage of learner maturity should be characterized by structured learning activities that allow for some input from the adult participants. Professional association meetings and conventions would be examples of adult education activities that combine structure with participant input. In this context, the participants are motivated to participate (or not participate) at their discretion. Convention schedules often offer a variety of concurrent sessions that allow adult learners to self-direct their learning to the degree that they can choose which sessions they want to attend.

Stage 3   *LOW PEDAGOGY, HIGH ANDRAGOGY*

Learners at this stage are considered self-directed to the point where they do not require structured learning environments or activities. The learners generally have extensive experience and can independently plan and implement their learning activities. The content and approach is completely at the discretion of the learner and participation is strictly voluntary. Adults have the option to choose whether or not to become involved in learning. Most of the self-help and liberal adult education programs offered through the public schools and/or community colleges should recognize that many of their adult participants may be classified as self-directed learners.

Stage 4   *LOW PEDAGOGY, LOW ANDRAGOGY*

Adult learning that occurs at this stage may be characterized as "happenstance." Adults acquire information or knowledge each day through a variety of channels including newspapers, magazines, radio, television, and conversations with other persons. Simply driving down a road we are constantly "learning" new information as a result of the messages we receive from our environment. In this stage, the learning that occurs is not the direct result of a highly planned or structured set of learning activities, nor does the learner necessarily have the desire or motivation to pursue what has been learned to any further degree. This stage could be characterized as "laissez-faire" learning and much of what is learned in this manner is quickly forgotten if not reinforced. However, occasionally, adults will learn something in their daily lives as a stage four learner that prompts them to move toward purposefully initiating an active learning process. Learning in this stage generally involves only the learner without intentional assistance from an adult educator or facilitator.

As adult learners mature, they may have less need (and desire) for highly structured learning activities; however, the need for support systems may continue to exist. Therefore, the role of the adult educator may also evolve in conjunction with the changing level of learner maturity of adult participants. The growth and maturity of adult learners, coinciding with the evolutionary role of the adult educator, requires flexibility among individuals, programs, and institutions. Adult education programs in the future will need to provide flexible support structures in order to respond to the changing needs and interests of adult learners.

Summary

Adult learning is a complex process that can be guided by a number of teaching and learning principles. Each principle should be considered in the context of a specific learning environment. Adult learning principles should not be viewed as universal truths, but rather as guidelines to inform the decision-making process involved in planning, conducting, and evaluating adult education programs.

CHAPTER 5

ADULT TEACHING METHODS*

Adults learn in a variety of ways. Therefore, it stands to reason that there are a variety of educational delivery formats and teaching methods that should be employed to facilitate the learning process of adults. This chapter examines a number of delivery formats and teaching methods that can be used effectively in adult education and describes some factors to consider in their selection and use.

Adult educators should realize that to effectively facilitate the learning process, learners must engage in activities that expand their knowledge base from what is known to encompass that which was previously unknown. To accomplish this process, adult educators should provide carefully planned learning opportunities for adults to expand their knowledge and skills. In selecting educational delivery formats and teaching methods, adult educators should strive to provide learning opportunities that produce the desired learning outcomes.

---

*Bryan L. Garton, Assistant Professor, Agricultural Education, University of Missouri–Columbia.

Educational Delivery Formats

One of the first decisions an educator will make is the selection of an appropriate educational delivery system. Common delivery systems include:

- Informal meetings
- Tours
- Formal courses (credit and non-credit)
- Workshops
- Institutes
- Seminars
- Conferences
- Conventions

Selecting the most appropriate educational delivery system is dependent on several factors. However, the delivery system may be prescribed by another individual or an organization, and not left to the discretion of the adult educator. If the decision is left to the adult educator, input from an advisory committee (see Chapter 6) is recommended. Factors to consider in selecting the most appropriate delivery system include:

1. Purpose of the educational program (program outcomes)

2. Objectives (learner goals/outcomes)

3. Anticipated number of participants

4. Participant travel (distance and expense)

5. Available facilities, equipment, and resources

6. Estimated budget (income and expenses)

After determining the delivery system, additional planning is necessary to conduct an effective adult education program. Specific details to be addressed relative to planning the adult education program are outlined in Chapter 8.

One important aspect of the planning process involves the selection of the teaching methods employed to fulfill the intended educational purpose of the program. The remainder of this chapter will be devoted to various teaching methods that can be used effectively in adult education.

Teaching Methods Used with Adults

Teaching methods can be divided into three major categories as follows:

1. One-way communication methods
2. Two-way (interactive) communication methods
3. Laboratory (skill development) methods

The selection of teaching methods from one or a combination of the categories is dependent upon the intended outcome (objective) and the maturity (readiness level) of the learners, from a knowledge and skill perspective.

One-way communication methods are most appropriate in situations where the objective is primarily focused on transmitting information from one or more sources to a group of learners. In this case, the intended audience would be assumed to have limited background information regarding the subject and would receive the greatest benefit by simply expanding their knowledge base through the acquisition of new information.

As adult learners gain in their educational maturity, due to an expanded knowledge base, utilization of two-way (interactive) communication methods may be employed to facilitate an exchange or dialogue between the information source and the adult learners. Adult educators should become adept at planning educational activities that combine methods, initially utilizing one-way communication methods and moving toward two-way (interactive) communication methods as learners "mature."

Laboratory teaching methods are used in learning situations where the objective is to gain or acquire knowledge and skill in the performance of a psychomotor task. Laboratory teaching methods are also useful in situations where observation of an application, practice, or skill is desired.

Selecting the Appropriate Teaching Method

When selecting the most appropriate teaching method(s), there are several factors that should be considered. Factors to consider include:

1. Objectives — desired learner outcomes
2. Subject matter (content)
3. Available facilities, equipment, and resources
4. Characteristics and backgrounds of the learners
5. Desired interaction of learners among themselves and with the instructor
6. Available time
7. Policies of the learning (educational) institution

The first factor to consider when selecting the most appropriate teaching method is the educational program's objective for teaching the subject (see Table 3). The instructor must determine if the purpose of instruction is to provide learners with new information (knowledge); teach learners how to apply new information (understanding); teach learners how to perform a skill; or help learners to modify, adopt, or clarify their attitudes or values. The objective is the most important factor in selecting appropriate teaching methods.

A closely related factor in selecting teaching methods is the subject matter (content) to be taught. Some topics naturally lend themselves to one-way communication methods because the content is new to the learners and they need a basic knowledge of the subject matter to internalize the information. As the learners mature in their knowledge of the subject matter, the teaching methods may change. Subject matter that involves learning a new skill or a new way of performing a skill, may be taught most effectively through demonstrations, role-plays, and computer-aided instruction.

The facilities, equipment, and resources available will greatly influence the teaching methods selected. Certain teaching methods may require particular room arrangements and space. Some teaching methods will require specialized laboratory equipment. In some situations, there may only be one piece of equipment to perform a demonstration, consequently limiting learners' opportunity to practice and acquire the skill.

TABLE 3.

Learner Goals and Applicable Teaching Methods

Learner Goals / Objectives	Applicable Teaching Methods
Knowledge: Goal is to gain an awareness of and internalize "new" information and make generalizations about experiences.	Lecture/presentation Panel discussion Visuals: films, slides, etc. Symposium Resource person (subject matter expert) Tours Field trips
Understanding: Goal is to gain an understanding of "new" information and apply the newly gained information to current problems or situations.	Group discussion Demonstration Problem-solving activity Case study
Skill attainment: Goal is to learn "new" ways of performing a skill through guided practice.	Demonstration Exercise/practicum Role-play Simulation Computer-aided instruction
Attitude/value modification: Goal is to modify or adopt "new" feelings or beliefs by experiencing greater success with the "new" vs. the "old" feelings or beliefs.	Panel discussion Group discussion Debate Role-play Case study Simulation Lecture/presentation

The learning characteristics, backgrounds, prior knowledge, and experiences of adult learners will affect which teaching methods to use. Also, the adult educator should determine if learners in the group have particular learning preferences. It is safe to assume that with any group of learners there will be a variety of learning preferences represented. Therefore, building a variety of teaching methods into the teaching-learning process would strengthen the learning experiences for all learners.

Closely related to the learning characteristics of adult learners is the amount of interaction desired. Interaction occurs in two forms: interaction between learners and interaction between learners and the instructor. This factor also corresponds to the desired outcome (objective) of the teaching-learning process. When the objective is to gain knowledge or understanding of the topic, we would expect to observe a limited amount of interaction. Conversely, when the objective is to gain proficiency in performing a skill or in attitude/value modification, we would expect to observe an increased amount of interaction in the teaching-learning process.

The amount of time available to teach a topic will also influence the teaching methods selected. Adult educators must plan ahead and anticipate the amount of time necessary to effectively teach a topic. Certain teaching methods, such as lecture/presentation, are well suited for providing a large amount of information in a short period of time. However, this may not be the most effective method of reaching the desired outcome (objective). For example, if the objective was for learners to perform a specific skill, then the demonstration teaching method might be more appropriate than lecture/presentation. However, using a demonstration may require a greater amount of instructional time. An adult educator must select the most appropriate teaching method(s) factoring in the time requirements needed to effectively use the various teaching methods.

The final factor to consider in selecting the most appropriate teaching method is the policies and regulations of the learning (educational) institution. Instructors are advised to check the policies and guidelines of their institution prior to incorporating certain teaching methods into the teaching-learning process.

Teaching Methods Using One-Way Communication

Lecture/presentation — The information is presented orally with a minimal amount of learner participation. The lecture/presentation teaching method is an efficient means of communicating factual information in a limited amount of time. This teaching method is also useful when the material is not readily available in other forms. Lecture/presentations can be enhanced and the amount of information that learners retain can be dramatically increased by the use of quality visual aids.

Resource person (subject matter expert) — An expert (consultant) provides an awareness of and new information regarding the subject matter. The primary purpose of using a resource person is to assist learning by providing experiences of the subject matter or information that is not available in other forms. A resource person will generally use the lecture/presentation teaching method followed with questions from the audience; however, other teaching methods could also be used.

Symposium — This consists of a group of brief presentations by resource persons on various aspects of the subject matter. Generally, there are from three to six presentations, each between 5 to 20 minutes in length. After the presentation, the presenters may participate in a panel discussion, question each other, or respond to questions from the audience.

Panel discussion — A panel of resource persons or a group of learners talk among themselves, present their ideas, and possibly come to some general agreements regarding the subject matter. In a panel discussion, only the panelists talk while the audience (learners) listen to the panelists. Modification of the panel discussion method may include having the panelists respond to questions from the audience (learners).

Computer-aided instruction — This is an interactive instructional technique in which a computer and a specialized computer program are used to present instructional material, monitor learning progress, and select additional instructional material based on learners' needs and progress.

Teaching Methods Using Two-Way (Interactive) Communication

Group discussion — The entire group of learners participate in a discussion for the purpose of sharing information regarding issues, problems, or questions of the subject matter and analyzing and evaluating the information to arrive at some general conclusions.

Case study — A detailed analysis focuses on a particular problem or issue of an individual, group, or organization. Case studies can serve several purposes. They can be useful in generating a discussion of an issue or problem, they can be used to provide relevance and meaning to the subject matter, and they are useful in introducing and leading learners into defining a problem that needs to be solved.

Problem solving — Learners are actively engaged in defining a problem that needs to be solved or a decision that needs to be made, in identifying factors relevant to solving the problem, in seeking data and information to solve the problem, in formulating and testing alternative solutions, and in arriving at a solution to the problem. Several problem-solving techniques have been used successfully by adult educators. These techniques include:

1. The "Forked-Road" decision — Learners identify and analyze factors necessary to make a decision between two possible alternatives to solve the problem.

2. The "Possibilities-Factors" decision — This is similar to the forked-road decision, but in this case learners make a decision between three or more possible alternatives to solve the problem.

3. The "Given the Effect, Find the Cause" problem — Learners are presented with a problem that has identifiable symptoms. Learners must determine the possible causes, analyze and evaluate facts related to each possible cause to determine the most likely cause, and identify alternatives for correcting the problem.

4. The "Situation-to-Be-Improved" problem — Learners are presented with a detailed situation in need of improvement. Learners must compare the current situation to the characteristics of an ideal situation and offer recommendations as to how to move the current situation toward the ideal.

Role-play — A group of learners act out a situation or an incident to portray a common human relationship problem. Role-plays are useful in adding relevance and meaning to the subject matter by introducing learners to common human-relations problems.

Brainstorming — This process encourages the creative generation of ideas regarding a specific topic in which learners contribute suggestions in a spontaneous and noncritical environment. Brainstorming activities are often used as a pre-planning activity to formulate ideas for future learning sessions.

Teaching Methods for Laboratory (Skill Attainment)

Demonstration — This method is useful in illustrating and explaining, in an orderly and detailed way, how to perform specific skills and procedures. Process demonstrations are very useful in (a) teaching psychomotor skills, (b) developing an understanding of how things operate and function, and (c) modeling how to perform new practices and/or procedures. Result demonstrations are used to illustrate the "outcomes" resulting from variations in a process.

Tour / field trip — In this activity, the learners travel to specific locations to learn the specified objective. Could include the observation of a situation, the observation of practices in action, or simply bringing the learners in contact with individuals and practices that could not be observed under existing instructional circumstances. Field trips are usually to one site or location, while tours may involve visiting several sites. Tours may be conducted within a single day or over several days, or weeks.

Using Multiple Teaching Methods

We know that adults, all learners for that matter, learn in a variety of ways. Furthermore, experience and research have indicated that when adult educators add variability to their teaching, learners' attention, motivation for learning, and actual learning are enhanced. One way to add variability to the teaching-learning process is through the use of multiple teaching methods in the teaching of a particular subject (topic).

Many adult educators will combine two or more teaching methods to more effectively communicate the subject. For example, the lecture/presentation method of teaching can be enhanced by adding group discussion at the appropriate time in the learning process. A role play could also be added during a lecture/presentation to add clarity to a key point being made.

TABLE 4.

Commonly Used Teaching Methods

Teaching Method	Purpose	Suggested Strategies for Success
Lecture/ presentation	To provide a large amount of information (knowledge) in a limited amount of time. An efficient method of providing information not readily available in print or other forms.	• Develop attention getters (interest approaches) to gain and hold learners attention. • Prepare a detailed outline with key points. • Organize and structure the material in a logical sequence for the learners. ⇨ Begin with an overview. ⇨ Build on the familiar before going to the unfamiliar. ⇨ Discuss events in chronological order. ⇨ Build from the simple to the complex. • Make frequent changes in the teaching-learning environment. • Frequently use visual aids to add clarity and variability. • Show enthusiasm.
Resource person (subject matter expert)	To provide knowledge and experience from a recognized expert on the topic.	• Provide resource person with: ⇨ Topic of discussion (Clearly communicate objective.) ⇨ Length of the session ⇨ Learners' prior knowledge and experience of the topic ⇨ Special circumstances ⇨ Number and background of learners • Determine the resource person's equipment needs. • Identify needs or special requests of the learners. • Prepare learners prior to the resources person's presentation. • Promote the raising of questions by the learners.

(Continued)

TABLE 4. (Continued)

Teaching Method	Purpose	Suggested Strategies for Success
Symposium	To provide a variety of viewpoints on a particular issue, problem, or topic from a panel of experts.	• Clearly define the issue, problem, or topic to be discussed, and share with panelists prior to the symposium. • Select four to six experts for the panel. • Select panelists with a diversity of backgrounds and experiences. • Have each panelist prepare a brief presentation on the issue, problem, or subject. • Have the facilitator make transitional comments between each presentation. • After the presentations, let the panelists participate in a panel discussion, question each other, or respond to audience questions. • Plan for a follow-up discussion during a subsequent session.
Panel discussion	To provide an opportunity for experts or a group of learners to present differing view points on a topic, issue, or problem to other panelists and the audience (learners). The discussion of the panel should stimulate the audience's thinking.	• Clearly define the issue or problem to be discussed. • Select (possibly from the learners) and prepare members of the panel. • Designate a leader for the panel. • Arrange the learning environment with the audience (learners) in mind. • Keep the discussion on-task and within a specified time frame.
Computer-aided instruction	To provide an opportunity for adults to learn the subject at their own pace. An effective method of providing active learning with immediate feedback and re-enforcement.	• Provide a learning environment where learners can work without distractions. • Provide up-to-date computer technology. • Preview software (tutorial) programs extensively. • Select software programs that provide active learning (interaction). • Select software programs that require a variety of learning preferences (auditory, visual, and tactile). • Select software programs that provide immediate feedback.

(Continued)

TABLE 4. (Continued)

Teaching Method	Purpose	Suggested Strategies for Success
Group discussion	To provide an opportunity for learners to think together constructively for purposes of learning, solving problems, making decisions, and/or improving human relationships.	• Guide learners into selecting the topic for discussion. • Prepare a list of leading questions that will stimulate thinking and discussion. • Arrange the learning environment to promote discussion. • Establish an atmosphere in which learners have an equal opportunity to participate.
Case study	To provide an account of an actual problem or situation that has been experienced by an individual or group. An effective method of provoking controversy and debate on issues for which definite conclusions do not exist.	• Present the case in writing with 3 or 4 questions that will generate discussion. • Be prepared with leading questions to stimulate thinking and discussion. • Arrange the learning environment to promote discussion. • Establish an atmosphere that promotes an equal opportunity to participate in the discussion. • Guide the discussion toward the intended outcome.

Problem solving	To provide the opportunity for learners to solve a problem through the collection, application, and assessment of information. An effective teaching method to encourage learners to inquire into, and think critically about, a topic.	• Follow a systematic procedure in solving a problem:
		  ⇨ Clearly formulate and define the problem.
		  ⇨ Identify the relevant factors pertaining to the problem.
		  ⇨ Collect the information needed to solve the problem.
		  ⇨ Examine possible solutions to the problem.
		  ⇨ Select a tentative solution and/or alternatives.
		  ⇨ Test proposed solutions.
		  ⇨ Evaluate the results, and determine whether future action is required.
		• Choose problems that are relevant to the lives of the learners.
		• Use a case study, role-play, or other teaching method to involve learners in defining the problem.
		• Make problem solving an active learning process that requires learners to inquire into the topic.
		• Let learners gather the information.
		• Have learners use a variety of sources to collect information.
		• Have learners summarize what was learned from solving the problem.

(Continued)

TABLE 4. (Continued)

Teaching Method	Purpose	Suggested Strategies for Success
Role-play	To provide learners with the opportunity to experience common human-relations problems in a secure environment.	• Determine the specific objective(s) to be accomplished. • Design the role-play to meet the specified objectives. • Prepare learners participating in the role-play for their roles, and provide situations and scripts if necessary. • Analyze and summarize the role-play to relate to the specified objective.

Brainstorming	To solicit creative ideas or to identify possible solutions to problems. Allows learners to express opinion and ideas without the threat of being judged by other learners.	• Begin brainstorming session with a specific topic or problem.
		• Have the facilitator explain the ground rules of a brainstorming session:
		  ⇨ All opinions and ideas pertaining to the topic are welcome.
		  ⇨ Judgment of opinions and ideas is not allowed.
		  ⇨ Criticism is not allowed.
		  ⇨ Expanding on the ideas of others is encouraged.
		• Emphasize that quantity of ideas is desirable; the more ideas offered, the better.
		• Have the facilitator keep the group focused on the topic or problem.
		• Have a recorder document all ideas.

Demonstration	To model the correct step-by-step procedures needed when performing a specified task.	• Keep the learners interested by involving them in the demonstration. • Assemble all equipment and supplies, and rehearse prior to the actual demonstration. • Keep the demonstration simple — do not try to teach too much in one demonstration. • Check for clarity by asking questions, and watch for signs of confusion. • Outline steps (procedures) using some form of visual aid. • Restate key points several times. • Provide time for learners to apply and practice the "newly" acquired skill.

Continued

TABLE 4. (Continued)

Teaching Method	Purpose	Suggested Strategies for Success
Tour/field trip	To provide an opportunity for learners to observe practices, problem situations, or to bring learners in contact with persons or objects that cannot be seen by other means.	• Determine the specific objective(s) to be accomplished. • Select an appropriate site. • Plan transportation (include maps if necessary). • Explain special circumstances of the site visit. • Plan a follow-up and summary. • Write "thank you" letters to the appropriate individuals.

While using the problem-solving method of teaching, an adult educator could potentially use a *case study* to lead learners into defining the problem; a *group discussion* to identify the factors relevant to solving the problem; a *resource person, field trip,* and/or *panel discussion* to seek the data and information needed to solve the problem; and a *group discussion* to arrive at the final solution to the problem. In this example, seven different teaching methods could be used, increasing the variability of the adult learner's involvement in the process. Whenever additional teaching methods can be used to add variability to the learning environment, effectiveness will be increased.

Summary

Selecting the appropriate teaching method is primarily dependent upon the objective of the adult education program. Teaching methods can be classified into three groups: instructional strategies that involve one-way communication, those that involve two-way communication, and those that involve laboratory activities. Using a combination of teaching methods will generally improve program effectiveness (as measured by attainment of the learning objectives) by accommodating the range of learning preferences represented in the target audience.

CHAPTER 6

ADVISORY GROUPS

The principles of adult learning described in Chapter 4 point to the need to involve adult participants in planning learning activities whenever possible. Frequently advisory groups have been utilized to provide input for planning adult education activities. Such groups can provide a tremendous boost to help ensure the success of the program.

Some of the advantages of using advisory groups in planning adult education programs include:

1. Broader range of input for program design and organization.

2. Vested interest developed among a significant part of target audience.

3. Variety of perspectives can be considered.

4. Source for potential recruitment and public relations efforts.

5. Assist with conducting and evaluating the program.

6. Assist with identifying resources needed for program operation.

7. Utilizes group dynamics and thought processes to design the program.

Although there are numerous advantages to using advisory groups in conjunction with adult education, there are some disadvantages that should be recognized. Most importantly, the time it takes to make decisions is usually increased. Adding steps and the number of people involved in the decision-making process increases the time it takes to arrive at a decision, proportionally, if not exponentially. Therefore, adult educators should be cognizant of areas in which advisory groups should be consulted (and their input utilized in the decision-making process) and areas where advisory group input may not be necessary or appropriate.

Advisory group input should be sought in areas where the decisions (and the resulting implications) would be better than if the decisions were made by a single individual. Although decision-making efficiency may be reduced when using advisory groups, adult educators should recognize the trade-off between efficiency and effectiveness regarding the quality of the decisions. Higher quality decisions generally result from using advisory groups in the planning process; it simply takes longer to get decisions made.

Using advisory groups supports the democratic philosophy of participant involvement in the educational planning process. Involving adults in planning educational programs (that will ultimately have an impact on their lives) creates an atmosphere of ownership and self-directedness. Adults who participate in the planning process generally develop a vested interest and will work more diligently to achieve program success.

Advisory committees have the potential to contribute to an overall public relations effort on behalf of the adult education program. Advisory group members should be viewed as the eyes and ears of the target population. Utilizing an advisory committee effectively will enable adult educators to gather ideas and perspectives from a range of individuals that assemble specifically for that purpose. In addition, advisory group members have the potential to take advantage of the "multiplier effect" in communicating with a broader range of constituent groups about the adult education program.

Two-way communication is vital for an advisory group to be effective and should be strongly encouraged. Advisory group members should be asked to bring their ideas, suggestions, and perceptions to advisory group meetings for consideration. In addition, advisory group members should be expected (and specifically asked) to promote the adult education program among their constituents in the target audience. These expectations suggest the need for advisory committee members who possess effective speaking, writing, and listening skills.

FIGURE 8.

**Advisory
Committee
Relationships**

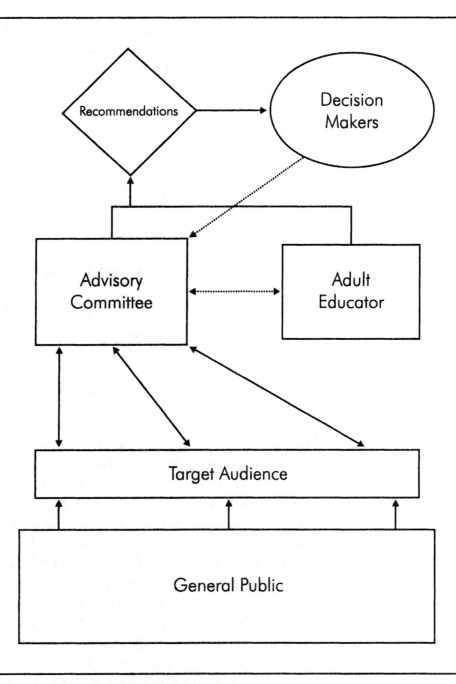

Member Characteristics and Selection

The effectiveness of adult education advisory groups can vary due
to a number of different factors. However, the characteristics of the

members that comprise the advisory group is an important consideration in the selection process. Effective advisory groups are usually comprised of individuals that:

1. Are respected among their peers

2. Reflect the diversity of the target audience (management, labor, age, gender, ethnicity, etc.)

3. Are knowledgeable of the industry/subject matter

4. Are open-minded and progressive

5. Are sincere, dedicated, conscientious, credible, principled

6. Speak their own mind, but can compromise

7. Are available/willing to meet and serve

Selecting appropriate advisory group members is critical to the success of the program. In some situations, adult educators can interview and "hand pick" members. Other situations may dictate the use of an intact group or existing committee to serve in an advisory role for the adult education program (e.g., Parent-Teachers Association, organization executive committee, board of directors, church council, etc.). It is recommended that an advisory committee consist of a representative mix by age, education level, race, gender, ethnicity, socio-economic status, management, and labor to provide the broadest range of input into the discussion. If possible, the adult educator should screen a list of potential advisory group members to identify those that possess the minimum characteristics outlined above and forward the list to an administrative unit for formal appointment to the advisory group. This procedure will increase the likelihood of success by adding significance (i.e., prestige) to the appointment. Furthermore, having advisory group members appointed by an administrative unit (school board, executive committee, director, president, superintendent, etc.) increases the likelihood that recommendations developed by the advisory group will be reviewed objectively. Advisory groups that consist of an intact group or are appointed solely by the adult educator, may be viewed as being biased and not serving in an officially recognized capacity. Therefore, whenever possible it is recommended that advisory group members be officially "appointed" by an administrative unit that has responsibility and authority over the adult education program.

Functions of Advisory Groups

The role of advisory groups should be to serve three major functions relative to adult education programs, namely: PLANNING, ORGANIZING, and EVALUATING. All these functions are important for an effective adult education program, and advisory groups can provide significant and unique contributions in each area.

PLANNING begins with the initial determination of needs. The process of needs assessment will be described in greater detail in Chapter 7. However, the advisory committee should have a significant role in assessing the educational needs of the target population. Advisory committee members should be involved in planning the overall needs assessment to determine:

1. What types of needs assessment information should be collected?

2. Who should be contacted to provide the information?

3. How should the information be collected?

4. What sources of information are currently available?

5. How will the needs assessment data be analyzed?

6. Who will interpret the needs assessment data?

Each question is an important consideration in conducting a comprehensive assessment of the educational needs of a target audience. Input from the advisory committee related to each question will provide a broader base of support in the overall program planning process. In practice, advisory committees may not be asked to address each and every question outlined; however, it is recommended they be involved in the needs assessment process at some level.

In addition to involvement in the needs assessment, the advisory committee should also provide advice regarding the scope, sequence, and structure of the information presented in the adult education program. After determining the needs of potential program participants, the advisory committee should be asked to assist in developing an overall outline of the content of the program and to provide input with regard to teaching/learning strategies. Specifically, the advisory committee should specify the overall purposes, goals, and objectives of the program. This information is critically important to guide the process of planning an effective adult education program.

After deciding on the purpose, goals, and objectives, the advisory committee may be asked to make recommendations regarding how each of the objectives should be fulfilled in the adult education setting. However, final decisions regarding specific instructional strategies are usually left to the discretion of the adult educator. Low cost, single session activities may not require extensive involvement of an advisory committee regarding the above planning considerations. However, many programs are more extensive and should seek advisory committee input regarding the decision points outlined.

ORGANIZING the adult education program is the second major function of an adult education program advisory committee. Advisory committee members, selected on the basis of their role and stature among the target audience, are well-suited to provide advice regarding the following program features:

1. Meeting time (day of week, time of day, season of year)

2. Number of sessions

3. Length of each session

4. Facilities/location (distance, handicap accessible, equipment needed, functional design)

5. Support services (copy machine, food service, restrooms, child care)

6. Publicity/promotion (mailing list, newsletters, flyers, advertisements)

7. Budget (fees, sponsorships, scholarships)

8. Certification/recognition (attendance, skills, development, skill achievement standards)

9. Resources (instructors, guest speakers, publications, equipment, supplies, materials)

Members of the advisory committee add a degree of legitimacy to decisions that need to be made regarding each of the program features.

enced members) should not be underestimated. Remember, advisory committee members serve a two-way communication function; they are expected to bring new ideas to the discussions and to take information they receive to other members of the community or target audience.

Leadership for the advisory committee should be provided by an experienced member. Adult educators often assume the leadership role; however, this practice is not recommended because of the tendency to stifle free and open discussion regarding issues affecting the program. Adult educators should serve as the secretary for the advisory committee to facilitate recording and distribution of the meeting minutes, sending notices of meetings with an agenda, and preparation of recommendations forwarded on behalf of the advisory committee.

Adult educators will often find that it takes more time to support, encourage, and prepare the elected chairperson of the advisory committee than it would have taken to fulfill the leadership role themselves. However, the long-run benefits of having the adult educator assume a less prominent role in leading the advisory committee will usually produce greater benefits. Advisory committee members will generally be more open to discussion and take their role more seriously when they recognize that leadership for the group's activities is provided by one of their peers.

Conducting an effective advisory committee requires adequate planning and preparation. Advance notification (at least one week) is necessary to allow advisory committee members to plan for their attendance and participation. Notice of upcoming meeting dates should be announced at each meeting and should be prominently noted in the minutes of each meeting. Therefore, minutes of the previous meeting should be mailed to all committee members within one week after the advisory committee meeting, rather than waiting to distribute them at the following meeting.

Notices of upcoming meetings should include a detailed agenda (if at all possible) and specifically request advisory committee members to collect information or reflect on specified issues to be presented and discussed. Members who attend advisory committee meetings without prior knowledge of what is to be presented and discussed will be less able to provide the thoughtful input needed for quality recommendations. Often, advisory committee members find it helpful to receive a personal phone call a day or two prior to the meeting encouraging their attendance. This direct and personal attention will frequently motivate members to attend and more actively contribute to the activities since they were "personally invited."

Meeting time and duration is also important. The starting time of the meeting should be clearly stated and followed. Busy people who are present at the designated time for the meeting to start should not be "penalized" by having to wait for late-comers to arrive. Being late is a habit for some people and if advisory committee meetings are allowed to start late, it becomes a self-fulfilling prophesy. Ultimately, other members will start arriving late anticipating the meeting will not start on time anyway.

Along with maintaining a designated starting time, it is suggested that an ending time be specified. The adage "The work will expand to the time allotted" is frequently true. Many people work more efficiently under a time constraint. Therefore, clear communication of the start time, stop time, and work to be completed will usually result in a more productive meeting. Development of a structured agenda, with approximate times for each item, although restrictive to some members, will usually help to keep the meeting focused on the priority issues. Specifying an ending time for the meeting also allows participants to plan other activities that may be scheduled after the meeting. It is important to keep in mind that effective advisory committee members are often busy with a number of activities. The more clearly expectations are communicated, the more likely those expectations will be achieved.

Printed materials, reports, and statistical data should be distributed well in advance of the meeting in which they will be discussed. Do not waste valuable meeting time asking members to read information. Maximize the use of the 1½ to 2 hours allocated to face-to-face meeting time to address significant issues. Advisory committee members will quickly become disenchanted with meetings if they feel their time is being wasted. Light refreshments may be provided during the meeting. However, a structured break time should be avoided for meetings of less than two hours in duration. Advisory committee members should be made to feel welcome to warm their coffee, use the restroom, etc., during the meeting, as long as they do not interrupt the flow of the meeting. Remember, adults do not want to be treated like children and have to wait for a recess to use the restroom or get a drink.

Adult educators must be cautious to avoid dominating the discussion or creating the appearance of "railroading" recommendations. Proper utilization of an advisory committee requires that the adult educator have faith in the advisory committee members to make the best decisions and recommendations possible. Therefore, the adult educator should make every effort to provide complete, accurate, and unbiased information available to the advisory committee for their deliberations. Advisory committees that deteriorate into a "rubber stamp" endorsement of ideas promoted by the adult educator soon lose their effectiveness in planning, conducting, and evaluating adult education programs.

Advisory committees should be provided with as much information as needed and available for their consideration of the issues under discussion. Allowing free and open discussion is also necessary to encourage input from the different member perspectives. Recommendations should be developed by the advisory committee members with minimal guidance and direction from the adult educator. Otherwise, the advisory committee may become viewed as a "pressure group" to promote ideas generated by the adult educator. Some administrators do not support the use of advisory committees. This lack of support is based on the potential of the advisory committee to overstep its role and become a pressure group to promote a specific agenda, rather than maintaining a focus on providing recommendations for program improvement.

Open communication between advisory committee members, administrators, and adult educators is needed. Administrators who perceive the advisory committee to be acting only to endorse recommendations and preferences of the adult educator will place less value on those proposals than if the suggestions were developed through a more open, democratic process.

Committee or Council

In many references, the terms *advisory committee* and *advisory council* have been used, at times, interchangeably. This chapter has consistently used the term *advisory committee* from the perspective that the term generally refers to an advisory group with its primary focus on a single program (e.g., nursing, welding, computer programming, business, etc.). Many of the concepts and suggestions previously presented are equally appropriate for an advisory council, which has the distinction of being a comprehensive advisory group for a broad range of programs. Therefore, an advisory council may be comprised of representatives of a number of program advisory committees in a given institution or organization. Advisory councils may also be comprised of individuals without any direct ties to a given program advisory committee but rather have been selected to represent a broader segment of a community or population. Advisory councils should also serve the three major functions of planning, organizing, and evaluating adult education programs, albeit on an administrative (or institutional) level rather than a programmatic level. Operating guidelines for advisory councils would be quite similar to those of advisory committees. The major distinction would be in the scope of the issues that are appropriate for each group to address.

Summary

Advisory committees serve a useful purpose in adult education programs. Through the use of advisory committees, adult educators can involve more people in the decision-making process relative to planning, organizing, and evaluating the program. This involvement adds to the quality and effectiveness of decisions made regarding the adult education program.

Advisory committees have the potential to significantly improve the quality of an adult education program. Conversely, a poorly run advisory committee can be divisive and lead to deterioration in the adult program. Therefore, it is imperative for adult educators to adopt guidelines related to operating procedures with advisory committees. These operating procedures will increase the likelihood of success with an advisory committee, although many other factors will also affect the committee.

7 NEEDS ASSESSMENT

Frequently, adult educators use the term *needs assessment* to identify problems or situations that could be solved or improved through educational activities. Previous authors have offered numerous definitions of the term *educational need*. One common thread among the definitions involves the gap or discrepancy between a current condition and a desired condition. Therefore, an educational need can be defined as the gap between a current level of knowledge or skill and some desired level of knowledge or skill (Knowles, 1980). More specifically, the term *educational need* encompasses the learning required to achieve a desired level of knowledge or skill in a learner.

FIGURE 9.

Illustration of Educational Need

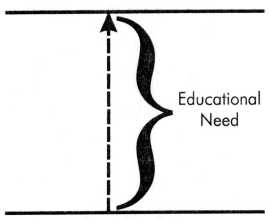

Desired Level of Knowledge/Skill

Educational Need

Current Level of Knowledge/Skill

Needs assessment can be viewed at the micro-level by examining a skill or knowledge deficit for an individual or group of individuals. At the macro-level, the gap between a current state and some desired state may be targeting an entire business, industry, community, etc. Although the steps and guidelines to follow in conducting a needs assessment are similar for the micro-level or macro-level, there is considerable difference in the types and sources of data that would be collected and examined in each context.

Venable and Mott (1995) identified four models that may be employed in completing a needs assessment. Many adult education programs have utilized various aspects of the four models to triangulate the results of the needs assessment efforts. Hybridizing the needs assessment for a specific adult education program is important to maximize the reliability and validity of the needs assessment data.

The four models identified by Venable and Mott (1995, pp. 44-49) were:

- Key Informant Model
- Community Forum Model
- Social Indicators Model
- Survey Model

A brief overview of each model will be presented in the following paragraphs; however, for a more complete description, the reader should refer to Venable and Mott (1995) and the additional references cited in their work.

The **Key Informant Model** can be a very efficient means of collecting needs assessment data. However, there are two implied assumptions that should be recognized before this model is adopted. First, the informants used to provide the needs assessment data are assumed to be knowledgeable of the "needs" of the target audience. Key informants should be stakeholders in the adult education program; however, they are not necessarily members of the target audience that would be potential participants in the adult education program. The second implicit assumption is that the needs identified by the key informants coincide with the views of the targeted program audience. The potential discrepancy between the needs identified by the key informants and the "felt needs" perceived by program participants has a direct effect on the success of the adult education program. Key informants who identify *needs* that do not coincide with the perceived needs (i.e., *wants*) of the

FIGURE 10.

Relationship Between Wants and Needs

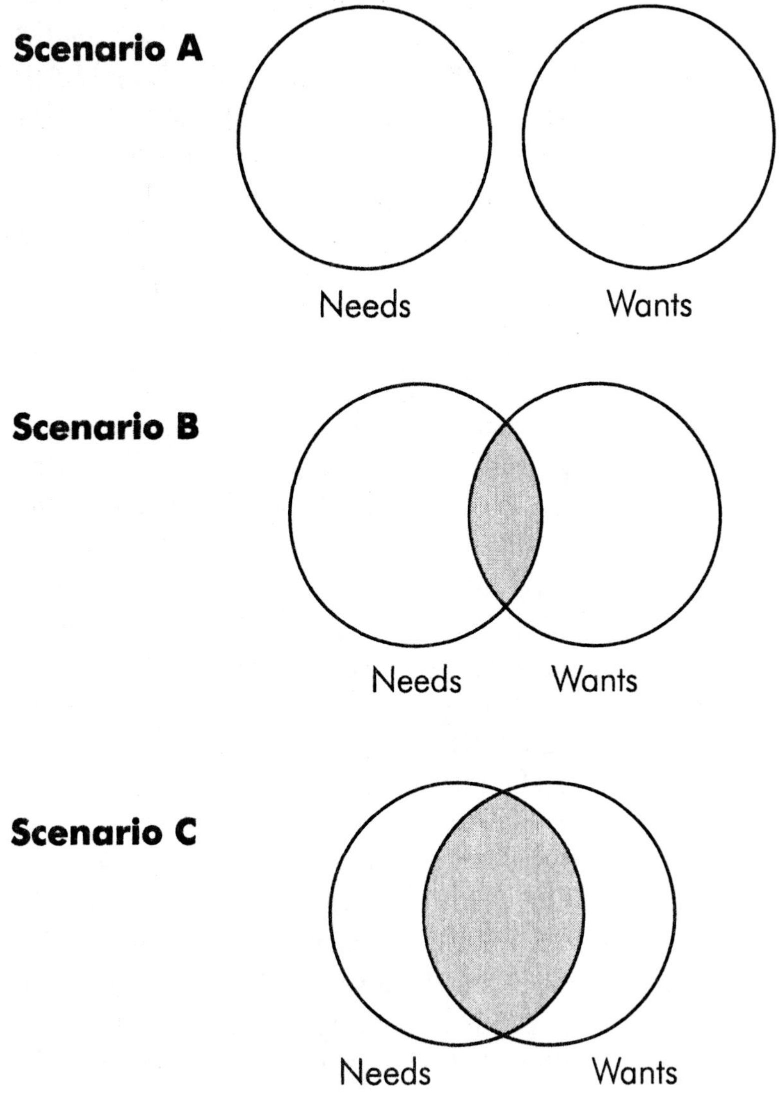

Scenario A

Needs   Wants

Scenario B

Needs   Wants

Scenario C

Needs   Wants

participants may result in low rates of participation or dissatisfaction with the program (see Figure 10).

In scenario A, there is no overlap between the *needs* perceived by key informants and the *wants* of program participants. This graphically illustrates a potential pitfall in using the key informants model as the sole source of needs assessment data. To secure attendance in a program developed under scenario A, some form of inducement (i.e., extrinsic motivation) would be necessary since the participants do not envision the program as providing anything they *want*.

Scenario B depicts a brighter outlook for an adult education program in that there is some degree of commonality between the *needs*

upon which the program is based and the *wants* of the potential participants. The degree of overlap (i.e., shaded area in common) is **significant** in that program planners should focus marketing efforts on promoting the knowledge, attitudes, and skills perceived to be both needed
and wanted. To be successful, adult education programs must begin at
the point of overlap as illustrated in scenario B and strive to expand the
wants of the participants to overlap into the *needs* area. The educational process frequently results in the participants' increased realiza-
tion of their *needs* (especially as originally perceived by third party, key
informants), ultimately expanding their *wants* to encompass their *needs*
as well.

Scenario C depicts an even more desirable situation in which the
key informants' perception of the educational needs of the target audience are more aligned with the *wants* of the potential participants. The
benefits of this situation should be obvious in that the participants recognize and desire the knowledge, attitudes, and skills incorporated in
the program since their *wants* coincide more closely with the *needs*
identified by the key informants. This scenario reflects the advantage of
using a key informant model in utilizing an efficient means of data collection while maintaining reliability and external validity.

In situations similar to the one depicted in scenario A, key informants may provide their perspective of educational needs; however, if
those views are not shared by the potential participants; attendance will
either be low or the participants will need to have some inducement
(reward or coercion) to prompt their attendance. The key informant
model can be a very efficient and effective method of obtaining needs
assessment data; however, it is frequently utilized in conjunction with
other methods (models) of assessment to increase the validity of the
results.

The **Community Forum Model** involves a process of gathering the
perceptions of targeted individuals or groups in a community or
organization. The forum should be structured to allow each of the participants to provide input regarding their perceptions of needs related
to a particular program, problem, or issue. This model is particularly
useful in prompting a free-flowing discussion that has the potential to
produce the most accurate picture of the felt needs of the target audience. However, without an effective moderator, the forum structure
may also result in a negative "gripe session" without producing the
information needed to guide the planning process for an effective adult
education program.

The community forum model also has the advantage of direct two-
way communication. Persons collecting needs assessment information

can ask questions and clarify viewpoints, which may not be possible with information gathered through a mailed survey or telephone interview.

A major limitation of the community forum model is the logistics of when and where to hold the meeting. Accessibility to all members of the target audience is ideal, but difficult to achieve. Multiple meetings (at various times during the day, days of the week, with child care provided, etc.) may be necessary to obtain a representative view of all segments of the target audience. Adult educators should recognize that meeting logistics frequently present barriers to participation that must be overcome in order to gain an accurate picture of the perceived needs of the target audience.

Adequate notification of the community forum meeting date, time, location, purpose, etc., is also important. In some cases, transportation may be needed for persons who desire to attend and participate in the meeting but do not have access due to transportation barriers. Notification may also be hampered due to communication barriers related to the media employed (newspaper, radio, television, posted flyers, direct mail, etc.); the message (purpose, role of participants, complete and accurate date, place, time, etc.); and the receptivity of receiver (language, trust, culture, etc.). Each of these factors should be considered in planning any adult education activity, whether a community forum needs assessment or an educational meeting. However, the accuracy of the information collected in a community forum is a function of who attends; therefore, maximizing the representativeness of the target population is vitally important.

Methods of collecting information during the community forum also have implications for the quality of the needs assessment data collected. Video and audio equipment can be very efficient in recording words that are spoken; however, often speakers will measure their words carefully when they recognize they are being recorded. Participants should be assured that their comments will be held in confidence and their identity will not be revealed without their permission. Even with such assurances, recording equipment may inhibit participants to the degree that a note-taker may be preferable to record key concepts that are presented without interfering with the free flow of ideas.

The moderator's role is also critical to the success of a community forum, especially when constituent groups have opposing viewpoints. Care must be taken to ensure that all participants are free to express their views without fear of reprisal or intimidation by persons from different perspectives. The community forum as a needs assessment opportunity should be structured similarly to a "brain storming" activity in which everyone has a right to express their views and no one's ideas should be discounted or discredited in the process. Flip charts or poster paper can be used effectively to illustrate how the thoughts and ideas presented by the participants are being received. Recording comments of participants provides visual feedback that the input received from participants is being "heard" and not just passing through a "nodding head." In their "needs assessment" efforts, politicians and government agencies often use public hearings to gather perspectives of stakeholders in their "needs assessment" efforts.

The **Social Indicators Model** utilizes secondary data sources in planning adult education programs. Increasing access to information data bases is expanding the opportunity to use secondary information sources in the needs assessment process. The exponential growth of the Internet has placed extraordinary volumes of information at our fingertips. However, the caution "user beware" should be exercised when using such data. Without knowing the purpose, methods, audience, and method of collection used to compile the data base, it may be easy to misinterpret or misapply the information.

Governmental agencies (health, education, labor, commerce, agriculture, etc.) frequently collect statistics that can be used to supplement an overall needs assessment effort. Using information from such sources may be helpful in explaining situational characteristics in a larger context. However, it should not be viewed as an adequate replacement for information collected directly from a more focused target audience. Newspapers, magazines, and industry newsletters may also offer sources of secondary information that can be used in the needs assessment process. However, readers should recognize that the information presented in such documents may be biased or invalid in a specific situation. Information collected from secondary sources may contain specific facts; however, it is generally not held to the same standard as court testimony, which requires "the truth, the whole truth, and nothing but the truth."

The validity of information obtained through secondary sources is a major concern with the Social Indicators Model. Although the information may be factual, the question remains with regard to its applicability to the target population. Determining the validity of the information with respect to the target population or local situation is an important prerequisite to consider before using secondary data sources in a needs assessment.

The **Survey Model** is frequently employed in situations in which the target audience is well-defined and accessible. The major advantage of this model is the fact that the needs assessment data is gathered directly from potential program participants. The two major assumptions underlying this model are (a) the participants surveyed have the ability to accurately assess their needs <u>and</u> (b) they *want* the information/skills that they identified as a *need*.

Survey instruments or interviews (personal, telephone, or electronic) can be specifically developed for the purpose of needs assessment in a given situation. Therefore, adult educators should develop the ability to construct a needs assessment instrument that will enable them to secure reliable and valid data.

A needs assessment survey will often have three or more components included in the design of the instrument. The first, and most important, component is questions asking respondents about their perception of their need to acquire information or skills on selected topics. Likert-type scales (with five response alternatives) are often employed to allow respondents to complete the survey quickly and conveniently. Response scales often use descriptors ranging from Strongly Agree to Strongly Disagree or from Absolutely Important to Absolutely Unimportant, with a neutral response category in the middle of the scale (although there is some disagreement about including a "neutral" response alternative). Examples of items that might be included in a needs assessment survey are shown in Figure 11.

After compiling the responses from a number of potential participants, average scores for each item should be calculated, and the topics should be arranged in rank order from highest to lowest need. Thereafter, the adult education program should be planned to address topics that were identified as highest priorities.

The second component commonly included in a needs assessment instrument involves items to assess the demographic characteristics of the respondents. This information is helpful in examining the characteristics of the individuals who provided responses in order to subdivide responses into logical groupings by age, type of position, geographic location, or some other classification variable. Demographic data is also useful in helping to interpret the data collected. Specifically, it may be helpful to determine if the respondents were representative of the target population or if only a select subgroup of the population provided feedback on the survey instrument. Either situation is important to establish a context for reviewing, summarizing, and interpreting the needs assessment data collected.

FIGURE 11.

Sample Needs Assessment Survey

Name _____

Address _____

City, State, ZIP _____

Home Phone _____ Work Phone _____

Job Title _____ Years in Current Position _____

I. Directions: Circle one response under the "Present Level" column and one response under the "Needed Level" column to indicate your personal assessment of the need for each of the following topics in performing your <u>work role</u>. (1 = None   2 = Some   3 = Average   4 = Above Average   5 = Excellent)

Topic	Present Level	Needed Level
Time Management	1 2 3 4 5	1 2 3 4 5
Departmental Budgeting	1 2 3 4 5	1 2 3 4 5
Cost Accounting	1 2 3 4 5	1 2 3 4 5
Performance Reporting	1 2 3 4 5	1 2 3 4 5
Strategic Planning	1 2 3 4 5	1 2 3 4 5
Employee Relations	1 2 3 4 5	1 2 3 4 5
Microcomputer Wordprocessing	1 2 3 4 5	1 2 3 4 5
Microcomputer Spreadsheets	1 2 3 4 5	1 2 3 4 5
Microcomputer Database Management	1 2 3 4 5	1 2 3 4 5
Using E-mail	1 2 3 4 5	1 2 3 4 5
Using the Internet	1 2 3 4 5	1 2 3 4 5

II. <u>Rank</u> the following topics in terms of your priority preference for inservice/continuing education.
(1 = Highest Priority; 6 = Lowest Priority)

_____ Leaderhip in the Workplace   _____ Handling Stress on the Job
_____ Management in Times of Change   _____ Balancing Work and Family
_____ Negotiating Peacefully   _____ Conflict Resolution

III. What other topics would you suggest for future inservice/continuing education programs?

_____

IV. Would you be interested in serving on a planning group for any of the above topics?
     o YES   o NO   If YES, specify topic: _____

Please return this completed form to: **Human Resource Services, Building A**

Thank You!

The third component often included in a needs assessment survey relates to information concerning the potential participants' ability to take part in the adult education program. Items may be included to determine preferences and/or barriers to participation related to topics such as:

- Location (building, distance from residence, etc.)
- Accessibility (facilities, equipment, handicaps, etc.)

- Start/end time
- Duration (number of sessions)
- Cost (registration, fees, supplies and materials, transportation, etc.)
- Availability of dependent care
- Day of week
- Season/month of year
- Instructor qualifications
- Interest in assisting with planning, conducting, evaluating, etc.

Open-ended questions may also be incorporated into the design of a needs assessment survey. Such questions often prompt the respondents to provide information about a topic or issue that is paramount in their mind at the time they complete the survey. However, difficulty may be encountered in compiling and summarizing the information collected in response to open-ended questions. It is usually helpful to develop a rubric to organize responses to open-ended questions into groupings to aid in identifying priority topics for adult education programs. However, using these types of questions in conjunction with other (i.e., objective items) can help adult educators identify important topics that may otherwise have been overlooked. It is recommended that both types of items (i.e., objective and open-ended) be included whenever possible.

Summary

The process of needs assessment is an important step in planning adult education programs. A variety of sources can be utilized to obtain information to be interpreted in the needs assessment process; however, direct input from the target audience is the most beneficial. Collecting information from a variety of sources increases the reliability and validity of the results of the needs assessment. Adult education programs that are well-planned and based on the results of a comprehensive needs assessment are more likely to be successful, than programs designed solely on the perspective of one individual. Adult educators should be constantly attuned to information and cues that provide indications of the needs and wants of the target audience.

CHAPTER 8

PLANNING ADULT PROGRAMS

There are many factors that must be considered in planning an adult learning activity. The word *planning* probably makes some people uneasy as they contemplate the time and effort needed to get the job done. It is easy to focus one's attention on all of the possible problems that might arise in conducting an adult education program. However, if a plan is properly developed, it can be a tremendous tool to ensure the success of the adult education program. Most successful adult educational programs have been well planned to meet the needs of the adults being served.

Planning Based on Program Objectives

When planning any type of adult education program, it is important to focus on the objective of the program. As discussed in previous chapters, needs must first be determined and then those needs must be translated into objectives. Program objectives are an important prerequisite to the planning process. However, the linkage between the program objectives and the means to achieve them is also critical. Persons involved in planning adult programs should ask themselves certain questions before beginning the planning process.

- What are the objectives of the program?

- What activities are needed to meet the objectives?

- Do certain steps have to be completed before others?

- What is the level of performance needed to achieve the program objectives?

Adult programs vary in complexity. Some programs are fairly straightforward and routine while others are more complicated. When the needs have been determined and translated into objectives, the focal point of the planning process has been identified.

Getting Help with the Plan

It is important to utilize input from a variety of perspectives in planning an adult education program. Planning adult education programs is not something that should be completed in isolation. If a program is to be successful, adult participants should be involved in the planning process to the extent possible. Learners generally perceive a program to be of greater value when they have had some input in planning the activities. Whether there is a program advisory committee or a temporary planning committee, the groups' contribution to the planning process will likely increase attendance and enhance a sense of shared responsibility for the program.

Planning

Planning adult education programs should include members of the target audience to help ensure that the interests of the participants will be met. Participants should be invited to participate in the earliest stages of the planning process. Previous chapters have already addressed the importance of advisory committees, assessing the needs of adults, adult characteristics, and methods of teaching adults. Later chapters will review concepts related to promotion and evaluation of adult educational programs. This chapter will focus on factors and components that should be considered in developing a comprehensive plan for evaluating adult education programs.

The adult education program plan should address all of the factors related to conducting a complete and successful adult educational program. The plan should include a complete outline, all resources, and a time line of all activities necessary to design and conduct the program in order to fulfill the objectives of the program. Although not all elements will be needed in every program, it may be helpful to use a checklist in establishing a complete plan. The following is a checklist of the major factors that should be considered in adult program planning:

- ☐ Content (program goals, program objectives, learning activities, subject matter, sequencing, etc.)

- ☐ Facilities (location, cost, availability, accessibility, seating arrangement, tables, A-V equipment, lighting, heat, air conditioning, fresh water, etc.)

- ☐ Logistics (starting/ending time, number of sessions, day of week, month of year, etc.)

- ☐ Resources (personnel, materials, consumable supplies, utilities, and equipment)

- ☐ Auxiliary Services (refreshments, parking and transportation, health and child care, lodging, copy/printing/binding/photo services, etc.)

- ☐ Administration (registration, financial management, safety, skill certification, CEUs, evaluation, and attendance)

[NOTE: See Appendix G for information related to the definition of CEU (Continuing Education Unit) and the criteria for awarding CEU credit for adult education programs.]

Content

The primary consideration in the overall adult education planning process involves decision making with regard to the program content. The initial steps in planning program content involves collecting and interpreting information as part of the needs assessment effort. Figure 12 (Translating Needs into Objectives, see Knowles, 1980, p. 125) illustrates how program content is influenced by the results of the needs assessment. As noted in Chapter 7, the needs assessment should involve input from several perspectives to triangulate the most accurate assessment of *needs*. Although the interests, goals, and desires of potential participants should be the primary focus, the subject matter should also be carefully examined to incorporate the range of knowledge, attitudes, and skills necessary for participants to learn the desired content (including performance standards related to professional licensure or certification). The third major source of needs assessment information relates to the "third party" perceptions of employers, leaders, politicians, and societal standards. The compilation of input from each of these three major sources constitutes a comprehensive needs assessment effort. However, the information compiled must be synthesized, analyzed, and interpreted before it can be used to guide and direct the program planning process.

Analyzing needs assessment information involves a three-step filtering process to focus on needs that are appropriate to address in the

FIGURE 12.

Translating Needs into Objectives

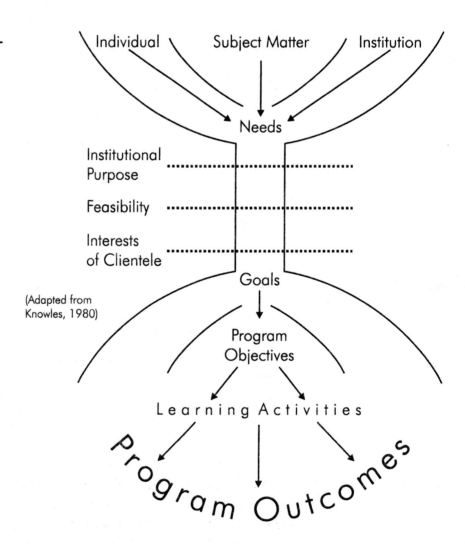

(Adapted from Knowles, 1980)

adult education program. Initially, the needs must be examined in the context of the "institutional purpose." Needs that fit within the mission, goals, and objectives of the sponsoring organization should be further considered as and addressed within the overall planning process. Needs that have been revealed, but do not appear to fall within the institution's purview (i.e., institutional role and responsibility) should be eliminated from further consideration. This filtering process is utilized as a screening device to prevent institutions from attempting to "be all things to all people." Institutions sponsoring adult education programs are better served by narrowing their focus of adult education to address needs that fit within a defined market niche.

The second filter to be applied as a screen for remaining needs is that of *feasibility*. This aspect considers issues of practicality regarding program planning decisions. Potential programs that are judged to be too costly, too risky, or too poorly attended may not be feasible to pursue. Therefore, program administrators should examine alternatives to identify and distinguish the feasibility of offering each program in the context of resource availability.

The third filter involves that of the clientele group or target audience. The level of interest within the target audience, as evidenced by the level of participation, will ultimately have an effect on adult education program offerings. Low attendance often leads to cancellation of adult education programs. This filter is often outside the control of adult educators in the planning process. However, it is a very real screening device that should not be overlooked.

Adult learner needs that successfully pass through each of the three filters should be considered in program planning. The first step in planning involves setting overall program goals. These statements consist of targeted outcomes the adult education program should be designed to accomplish. Program goals tend to be relatively broad statements that provide direction for the program, but without much specificity in terms of program content.

After the program goals have been established, the next step is to develop program objectives. These statements are more specific and measurable than program goals. Program objectives help to provide focus for program content and should relate to specific "changes in behavior" (see Appendix F — Levels of Learning Within Domains) expected of participants who successfully complete the program. Program objectives should be *written* in behavioral terms beginning with an action verb (e.g. describe, compute, create, design, evaluate, etc.) An example of a program goal and program objective for an adult education microcomputer program might be as follows:

Program Goal:	Learn how to use a microcomputer in an office setting.
Program Objective:	Create a form letter using a microcomputer wordprocessing program.

After the program objectives have been specified, the adult educator should begin to analyze the specific content knowledge, attitudes, and skills needed to achieve each objective. Organizing program content around the program objectives should proceed in a logical and sequential fashion. Arranging the content and learning activity sequence should take into account the learning style preferences of the target audience. Alternative content sequencing strategies may include:

- Simple to complex

- Chronological order

- Specific examples to general principles (inductive approach)

- General principles to specific examples (deductive approach)

Once the program content has been outlined, the next step in planning involves the selection of learning activities that can be used to facilitate the adults learning the desired knowledge, attitudes, and skills. Teaching methods used in adult education were presented in Chapter 5. Therefore, the important factor to re-emphasize at this point in the planning process is the need to focus on the program objectives. Generally, the action verb specified in the program objective will be a strong indicator of the teaching method that can best be used to achieve the stated objectives.

Facilities

In some cases, the choice of facilities may be limited, but in other situations, (e.g., selecting a conference hotel for a professional conference) there may be a number of different alternatives to select from. If possible, the program needs should be the guiding force to determine the facility selection, rather than having the facility dictate the program activities.

The content and learning activities, the educational aids, and the overall structure of the program must be considered when planning for facility use. Programs that require specialized equipment, such as computers, machines, tools, or worktables, should be sure to schedule the equipment and the facility that meets the needs and size of the program. Adult education programs that involve group discussion, may require a small, quiet room in which the chairs can be moved into a circle so participants can face one another. Consideration should also be given to whether the site is accessible to handicapped participants. The site should have ample, safe parking, and be within a reasonable proximity of participants so travel does not present a barrier to attendance. It is very important to remember that the facility should be comfortable and meet the overall program objectives.

Room arrangement is also an important consideration in program planning, and the type of arrangement depends on the objectives of the program. The primary consideration is what the participants will be expected to do during the program. Will the participants be expected to take notes or to use printed materials? If so, then tables and chairs may be needed. Will participants be facing the instructor, or will the room be arranged with chairs in a circle to facilitate discussion? Can the tables and chairs be moved to allow shifting between large group and small group activities? Will the educational personnel be up front or circulating among the audience?

In many cases, small groups become the primary venue for adult learning. There may be times when there is a large number of participants, and it may be desirable to divide into smaller groups during certain parts of the program. If learning depends on interaction among the participants, groups should probably be no larger than eight. There may be a need to create a more relaxed structure at the beginning of a program to break the ice and make participants feel more comfortable. This can create an atmosphere that encourages participants to take a more active role in discussions. There also may be times when more structure is needed to reach the program objectives. The beginning of the session may use a more structured arrangement to review certain information and then modify the room arrangement to facilitate discussion. The important point is to implement the appropriate room arrangement best suited for the type of learning activity being planned. The room arrangement can have a positive impact on the program by making participants feel at ease, creating interest, and encouraging participation.

Time

The frequency of the program and the amount of time for each session will vary depending on the program and the participants. Some programs will require extended periods of time for completion. Multi-session programs that involve weekly sessions should be planned during periods of the year when other time commitments among the target audience are minimal. Single topic programs, meetings, or discussion groups may involve only one or a few sessions to complete. Once a week is generally the best schedule for such programs. This will help keep the topic fresh and interesting for the participants. The length of each session in adult programs varies with the type of program being offered. For most evening sessions, 1½ to 2 hours is a standard length of time for most formal programs. Sessions up to three hours in length may be appropriate for skill development or workshop sessions in which startup and cleanup time must be considered. If the objective of a session requires more than two hours, break periods should be scheduled every 1½ to 2 hours to accommodate the attention span of most adults.

The time of the year, the month, and the day of the week can make a major difference in the success of an adult education program. For example, programs involving agricultural topics should not be scheduled to conflict with planting or harvest seasons. Scheduling almost any adult education program within two weeks of the Christmas holiday season would also be likely to yield disappointing results, unless the topic dealt directly with the holiday season in some way. Programs that involve seasonal topics should be planned well in advance so information from the program can be put into immediate use by the participants. For example, a personal income tax seminar in April may not produce good results, whereas a session scheduled in January might stimulate a lot of interest.

Community activities should be identified and considered to make sure the program does not compete with other scheduled activities for attendance. The advisory committee can be very helpful in selecting the best time of the year as well as the right day of the week to schedule an adult education program. The first step should be to identify several alternative dates and ask the advisory committee to select the date with the fewest conflicts. Unfortunately, planning a date for the program cannot accommodate everyone's schedule. The final schedule should be determined by the advisory committee whenever possible.

Resources

One of the major areas of concern in planning adult education programs is anticipating the type and quantity of resources necessary for a successful program. The primary categories of resources include personnel, materials, consumable supplies, and equipment.

The category of personnel includes instructors, resource persons, technicians, and evaluators. Although most programs will not involve personnel in all areas, it is important to plan for personnel who will be involved in the program. Persons participating in the adult education program need to be invited and scheduled well in advance to ensure their availability. Planning may also involve assigning responsibility to individuals within the advisory committee to provide the communication linkage with external personnel participating in the program. Double-checking arrangements and schedules a few days before the session is advisable to avoid problems with miscommunication.

Resource materials needed in the program may include textbooks, workbooks, pamphlets, manuals, guide sheets, problem sets, software, and a wide assortment of audio-visual materials. In some situations, the materials need to be purchased on an individual basis by the participants. In other cases, the materials need to be secured and available for use by the participants or the adult educator. Arranging for the availability of resource materials needs to be completed well in advance of the start of the program and often involves calculating a "rough estimate" of the number of participants expected. Scheduling the availability of loaned or rental audio-visual resources also needs to be planned in conjunction with the overall scope, content, and sequence of the program.

Ordering the consumable supplies needed for an adult education activity is an inexact science at best. Sufficient quantities of consumable supplies should be ordered to provide *at least* two sets of materials for each participant. It is important that sufficient supplies be available to allow each person to experience success in the program. Providing at least two sets of consumable supplies will allow each participant the opportunity to repeat the activity (at least once) in case they make a mistake on their first attempt. Adult educators should strive to make sure that adult participants experience success the final time they attempt an activity before moving on or concluding the program.

Scheduling the availability of equipment for use in adult education is very important in the overall planning process. Sophisticated, technical equipment with limited availability may require that program participants travel to a site where the equipment is located. It may be necessary to reserve a microcomputer laboratory for some sessions to allow participants to develop their skills in using software packages. Audio-visual equipment is often a major concern when planning convention sessions at hotels due to the scheduling and expense involved. It is especially important to identify the resource needs of guest speakers and resource persons in advance to ensure the proper equipment is available when needed.

Equipment that should be considered in the planning process includes chalkboards, overhead projectors, projector screens, video projector and monitor, television, speaker telephones, copy machines, microcomputers, printers, FAX machines, scanners, modems, etc. Each items should be scheduled for the time needed and in the quantities necessary to conduct a successful program.

Auxiliary Services

Planning for auxiliary services that may be needed to support an adult education program can encompass a broad range of areas. Each of the areas identified may be a consideration, however, only the most comprehensive adult education programs would attempt to accommodate participant needs in each category.

Adult education programs are not only educational activities, but also social gatherings. Refreshments for adult programs are usually desirable. Having coffee (both regular and decaffeinated), doughnuts or cookies, and possibly juice or soft drinks, is probably advisable depending on the participants and the time of the day. The advisory committee or planning committee can be a valuable asset in helping identify appropriate refreshments. If resources are not readily available to provide refreshments, it may be necessary to ask participants to supply their own refreshments. If the program involves health groups, public health topics, or adults that are health conscious, other types of snacks may be preferred. It is often wise to have coffee ready for those who arrive early. Refreshments may also be planned during scheduled breaks as well as after the formal program. Providing refreshments during and after the program may help members of the group become better acquainted and feel more relaxed. Participants that are more relaxed will generally contribute more during group discussions, which can lead to a better learning experience for all participants.

In addition to providing refreshments, some adult education programs will involve meal functions. Planning for group meals should take into account the dietary preferences of the participants in addition to the cost of the meal. Formal dinners that involve served meals are generally more expensive than buffet arrangements. Most food service planners will have a range of alternatives from which to select. Food service may also be arranged through vending machine and canteen service providers, depending on the location, needs, and wishes of the participants.

Parking and transportation availability can have a significant impact on the success of an adult education program. Most importantly, participants may "perceive" parking and transportation problems as insurmountable barriers to participation. Reasonable arrangements should be made and clearly communicated to the target audience to circumvent perceived problems with parking and transportation. In some cases, the location of the meeting place may need to be changed to provide better parking and transportation services and to overcome this potential barrier to participation. Accessibility for handicapped participants is also an important factor for the program planner to consider.

The participation of some members of the target audience may be limited without the availability of child care services. Knowledge of the target audience is extremely important to identify the need for providing child care. Again, input from an advisory committee can be helpful in determining if there is sufficient need to provide child care services in support of the adult education program. If the decision is made to provide child care service, the next consideration will be to determine whether the costs for the services should be borne by the participants or by some other source. The latter decision has important implications for adult education participants on limited incomes and will significantly affect participation by some members of the target audience.

Programs that require overnight lodging accommodations pose a unique challenge to adult education program planners. Usually, it is advisable to provide a variety of lodging options (especially with a range of prices) and allow the participants to make their own choice of lodging preference. However, if lodging will be provided, there may be less flexibility that can be extended to the participants. Generally, the participants should be allowed to choose if they prefer single or double occupancy in their lodging accommodations.

Other services that may be considered in planning adult education programs included access to public telephones, availability of FAX, e-mail, postal, photocopying, photo developing, binding, recreation, and publicity services. Each of these options pose unique problems and opportunities that can influence the success of an adult education program.

Administration

Administering an adult education program involves a wide variety of activities that may be performed by a single person in a small program, or a team of professionals in large-scale convention activities. Handling participant registration is a major undertaking for many adult education programs. It involves collecting information and registration fees, as well as the distribution of information, schedules, and other program documents. Early registration allows for program information to be mailed to participants in advance if that is desirable. On-site registration should be well-planned and orchestrated to avoid a chaotic situation in which participants congregate in a congested location without sufficient knowledge of the program. Nametags and other registration materials (including receipts if appropriate) should be printed in advance and compiled into packets that can be quickly distributed to participants as they arrive. The program schedule should be clearly outlined in the registration materials to communicate the program options available to the participants.

Funding and financial management are responsibilities that many adult educators do not feel prepared to fulfill. It is advisable to develop a well-documented plan for receiving and disbursing funds as part of the adult education program. Registration and fee payment records should be accurately maintained along with a record of any payment (including receipts) for any expenses associated with the program. A final report of all receipts and expenses should be prepared and audited as necessary to provide complete fiscal accountability for the financial management of the program.

In addition to registration and financial management responsibilities, some adult education programs involve assessment of skill development for professional licensure or certification, awarding CEU credit, evaluation activities, attendance records, and safety records. Each of these components should be considered in developing an overall program plan, including identification of the individuals responsible and a strategy for addressing each component.

Summary

Planning adult education programs is a complex process that is vital to the success of the program. Major components that need to be addressed in the planning process include content, facilities, logistics, resources, auxiliary services, and administration. Planning should involve input from an advisory committee whenever possible and take into account the needs and interests of the target audience.

CHAPTER 9

CONDUCTING ADULT PROGRAMS*

Previous chapters of this text have addressed numerous factors that should be considered in planning and organizing an adult education program. Chapters 1 and 2 laid the historical background and need for adult programs. Chapters 3 and 4 focused on the characteristics of adult learners and underlying principles of adult learning. Chapter 5 provided extensive examples of adult teaching methods, and Chapter 6 focused on the need to utilize an advisory committee as an integral and essential part of adult programs. Assessing the need for adult programs was the focus of Chapter 7, and Chapter 8 described how to translate needs assessment information into a plan for an adult education program. With this background in mind, it would be very easy to conclude there was nothing more to say about conducting adult education programs . . . just do it. However, many issues still need to be addressed if adult programs are to be successful. This chapter will focus on several issues, including putting the plan into action, working with budgets, scheduling facilities and personnel, working with other staff, connecting with industry, planning for professional growth and development, and balancing work and family.

---

*Steven R. Harbstreit, Associate Professor, Agricultural Education, Kansas State University, Manhattan.

Putting the Plan into Action

Following the suggestions outlined in the previous chapters should result in a plan for conducting the adult education program. Developing the plan is one thing, putting the plan into action can be quite another. Chapter 8 suggested the following issues to be addressed in the plan:

1. What type of facilities will be needed for the program?

2. What time the program begins and how long will each session last?

3. What dates will the program start and finish?

4. What will be the arrangement of the classroom or meeting room?

5. What resource persons will be needed and for what part of the program will they be needed?

6. How will the participants be involved? Will they be a resource for their own learning?

7. What materials, supplies, equipment and teaching aids will be needed for the program?

8. What type of refreshments will be needed and how will they be provided?

9. How will the program be promoted?

Each of these items need to be accomplished well in advance of the program. It is often the responsibility of the adult educator to work cooperatively with a number of other individuals to see that these issues have been addressed. The following are some helpful tips in dealing with some of these issues.

Scheduling Facilities and Personnel

Educators who work in secondary or postsecondary education settings may assume the facilities they use with their day school programs are available for use with an adult program in the evening. This assumption can cause numerous problems with administrators and other persons working with adult programs. It is important to schedule the dates and times for using facilities through whomever is responsible for coordinating facility use. With the proliferation of adult programs, it is not wise to assume that facilities will be available at any time you want them. Failure to reserve facilities can result in two or more groups meeting at the same time and place, expecting to begin their programs. This can result in confusion, delays, and some very unhappy instructors and participants.

Arranging meeting rooms can be a special problem for adult educators. Carefully select the location for your program to adequately provide for the comfort of adults learners. Many adult programs involve rearranging the room to facilitate special learning activities. Remember to take a few minutes at the end of the program to return the room to its original configuration. This small courtesy will endear you to others who use the facility and will promote the continued availability of the facilities needed to conduct the program.

If additional personnel are needed to conduct the program adequately, be sure to schedule them well in advance. If additional custodial services are required, this too should be arranged and in some cases budgeted as a program expense.

Working with Budgets

In most cases, when adult programs are conducted, finances are involved in some way. Most adult programs must "pay their own way." Therefore, careful budgeting is essential for successful and continued operation. Do not assume anything when preparing a budget for an adult education program. Some items that should be considered include:

1. What are the costs for providing the instruction, i.e., fee to be paid for the instructor or outside persons needed to effectively conduct the program?

2. What are the costs for using the facilities?

3. What supplies are needed, how will they be secured, and how will they be paid for?

4. Are there any fees that need to be paid to other organizations, i.e., memberships in state or national organizations with which your local program is affiliated.

5. Are there costs for field trips or other special activities that need to be a part of the budget?

6. What is the expected enrollment? Is there a limit to the number that can adequately be served in this type of course (as determined, for example, by the availability of appropriate equipment in a computer course)?

7. Will there be any costs associated with the promotion or evaluation of the program?

All of these items need to be addressed and the costs determined before contracts are signed with the instructor or promotional materials are produced and distributed. Nothing will upset program administrators more than insufficient budget planning for the program. Careful planning will not prevent any unforeseen expenses from occurring, but with proper budgetary planning the adverse effects should be minimized. One suggestion would be to have an incentive clause in the adult educator's contract that would provide a base salary for a minimum enrollment threshold, with additional salary based on participation rates above the threshold level. This practice encourages the adult instructor to take a more active role in promoting the adult education program.

Working with Other Staff

Effective adult programs involve interactions with a variety of individuals. Many times, colleagues and other staff persons can provide tremendous help, advice, and support for adult programs. It is important that their involvement be viewed positively and as a collaborative effort. Collaboration is the key to success for many adult education programs, and for long term sustainability, mutual benefits should result from the relationship. The following factors should be considered when working with other staff members:

1. What areas of expertise do other staff members have that are needed in the adult education program?

2. When are the staff members most available to provide assistance?

3. What personal/family issues should be taken into consideration before asking other staff to participate?

4. What can you offer that would be an incentive for staff members to participate in the adult education program?

5. What can you or your adult program participants offer that could benefit the staff persons in their work?

After considering these factors, you should contact the staff person to determine if they would be willing to help/assist with the adult education program. Remember, not all staff members are willing or able to participate. You should not perceive their refusal as a rejection of you or your program. Reminding them of the benefits they could derive from participation is always a good idea.

Connecting with Industry

Forming partnerships with industry is critical to the success of many adult education programs. Having access to cutting edge technology and the latest information is not only beneficial to the image of the adult education program, but in many cases essential for the success of the adult participants. Many participants may be depending on the adult program to upgrade their job skills or gain new skills so they can earn a living for themselves and their families.

The rate of change and the proliferation of new information makes it necessary to develop many industry linkages. As suggested in Chapter 6, using an advisory committee helps to ensure that the adult education program is both relevant and current. Members of the advisory committee should be viewed as the starting point for developing industry linkages. You should utilize them as contacts and resources for developing a network of additional industry linkages.

Personal contact is the best method to initiate a linkage with industry personnel. Your first contact, however, should not involve a solicitation of resources for your program. Most businesses are approached regularly to donate money or equipment to support programs or activities in the community. Since there are limits to the available resources, businesses must make choices concerning the programs they will support. The first contact should be informative, explaining the nature and objectives of your adult program. Encouraging businesses to participate as members of the program is a good way to "break the ice." Once the benefits of the program are outlined, your industry contacts may suggest ways they can provide additional support. Cultivating these linkages takes time and effort, but most of all patience.

Planning for Professional Growth and Development

The long-term success of an adult education program depends on well prepared, competent adult educators to provide program instruction and leadership. To maintain a high level of professional competence, adult educators should develop a plan for their own professional growth and development. Staying "current" is essential if adult education programs are to be perceived as viable opportunities for adult learning.

Adult educators should be encouraged to participate in state or national conferences and workshops related to their instructional responsibilities. Participating in such programs can boost the morale and enthusiasm of instructors working with adults. Adult education supervisors and program administrators should support participation by paying for the conference or workshop registration or by giving time off with pay to attend the conference or workshop. Regardless of the form, the perception of administrative support is essential. Administrators of adult education programs should have an overall plan for the continued support and development of adult educators. Since they come from a variety of backgrounds, it is essential to provide for their continued development both in teaching and in technical content areas. If we believe in lifelong learning for adults, this must include adult educators who are responsible for delivering instruction.

Balancing Work and Family

One area of an adult educator's life often overlooked is the necessity to balance work with family needs and obligations. Since adult programs are scheduled outside of the "9 to 5" work day, adult educators are often gone when other family members are at home. The extraordinary time commitments on adult educators should be acknowledged and compensated whenever possible.

Participants in adult programs are not usually involved every evening all year around, so the time away from their families is more limited. Adult program participants are usually willing to reschedule family activities or miss class sessions on occasions when they need to participate in family activities. They may even choose not to participate in an adult education program based on the time frame it is offered and the potential conflicts with family activities.

Adult educators, on the other hand, do not have the luxury of rescheduling or missing a program session because of family activities. Since they are responsible for the instruction, they must be present for the effective delivery of the program. To ensure that an adult educator can balance work and family responsibilities, he or she should consider the following:

1. Find ways to involve the family in some of the activities of the adult program. If there is a special activity to recognize program accomplishments (e.g., a banquet, a field trip, etc.), make it a family event. Adult participants often enjoy the opportunity to include their families in these activities as well.

2. If a family member has expertise that is appropriate to share and it would benefit the instruction being provided in the adult program, include that person as part of the instructional team. He or she will have a greater appreciation for the time and energy needed to carry on effective instruction for adults.

3. Consider upcoming family events/activities before scheduling adult program activities. Adult participants consider family commitments before deciding to participate, and the adult educator should consider them when planning programs and activities for which he or she is responsible. Not every scheduling conflict can be avoided. However, with careful planning, conflicts can be kept to a minimum.

4. Avoid over-scheduling. Some adult educators become workaholics and find great joy and pleasure in working with adults. Most adults understand that an educator has a life outside the adult program. Common sense and atten-

tion to time management will allow a balance of family activities and adult educator responsibilities.

Summary

Conducting effective adult programs involves a variety of activities beyond needs assessment, utilizing advisory committees, and planning the adult program. Putting the plan into action requires attention to budgeting details, scheduling facilities and personnel, working with support staff, connecting with industry, professional growth and development, and balancing work and family.

CHAPTER
10

EVALUATING ADULT PROGRAMS

Evaluation of adult education programs should be an ongoing process. Evaluation activities should be planned before the program begins (pre-formative, needs assessment), during the program (formative), at the end of the program (summative), and for a period after the program has been completed (follow-up). Before the program begins, adult educators need to plan how evaluation information will be collected to assure that program objectives are met. During the program, evaluation activities should be planned to collect information that can be used to make mid-program changes in order to meet the needs of the participants. Evaluation efforts at the end of the program should provide information about the success of the current program in addition to identifying changes that should be made in future programs. Evaluation efforts conducted well after the end of the program can be helpful to determine the long-term effects on participant knowledge, attitudes, and skills and their satisfaction with the program.

Focus on Improvement

Program evaluation involves the process of collecting and interpreting information that can be used to judge the quality and effectiveness of the program in order to make informed decisions. The primary focus of all evaluation efforts should be to formulate recommendations for program improvement. Although evaluation often requires making a judgment about the quality of an adult education program, the real benefit of program evaluation is the degree to which future programs can be improved. Adult educators should embrace program evaluation efforts from the perspective that the results can be used to provide guidance and information for adult education programs in the future.

The strengths and weaknesses of adult education programs should be identified and analyzed as part of the evaluation process. Program strengths should be reviewed and promoted in relation to their contribution to overall program success. Program weaknesses should also be examined to identify steps that should be taken to overcome the problem. Addressing weaknesses in adult education programs may be uncomfortable for some adult educators. However, the process should be viewed as an opportunity to learn from their experience, rather than a criticism of their performance. Recommendations should be formulated with regard to each weakness identified with specific steps to improve the situation. It should not be construed as a "cardinal sin" to identify weaknesses in an adult education program. The "sin" is in repeating the same mistake in later programs. Recognizing that a problem or weakness exists in a program is only the first step in the evaluation process. The second and more important step is taking action to correct the situation, thereby making improvements in the program.

Adult education program evaluation should also address issues related to effectiveness. One primary question that should be answered through the evaluation effort is "Was the program successful in achieving the objectives?" To answer this question, evaluative data needs to be collected that provides a measure or assessment of the degree to which each program objective was fulfilled. Once again, it is vitally important that program objectives be carefully and thoughtfully developed due to the pervading influence that objectives have on all other aspects of the program. In many instances, it may not be possible to collect direct assessment information relative to each stated objective. Therefore, indirect measures may be needed to collect evaluative information. Often, useful evaluation information can be obtained through informal conversations between and among participants before and after sessions, or during breaks in the program. Adult educators should be attuned to opportunities to collect feedback that may be informative in the overall evaluation process.

Formal, written evaluation forms are frequently used to collect information from participants regarding their assessment of the program. This information should be collected and interpreted carefully to ensure its validity. Quite often, three primary questions form the basis for such evaluation efforts, namely: (a) What did you like about the program?, (b) What did you not like about the program?, and (c) What suggestions do you have to improve this program in the future?

Purposes of Program Evaluation

There are two primary purposes to be served in conducting a program evaluation—accountability and decision making.

Accountability

Needs assessment is one form of program evaluation that addresses accountability. A needs assessment is primarily performed to provide input and guidance into the program planning process in order to formulate appropriate program goals and objectives. This process, by its very nature, increases the accountability of the program due to the fact that the "need" for the program has been validated. Programs that are not based on needs assessment information may have accountability problems if the goals and objectives of the program are not recognized as local needs.

The *results* of the adult education program should also be examined in the evaluation process. Program outcomes should be reviewed with respect to the stated program objectives. Again, the importance of the objectives should not be underestimated. It is the objective that provides the direction for the program from the initial needs assessment through the final evaluation. The primary criterion question to be answered in evaluating the results of an adult education program should be: "Did the program achieve the stated objective?"

Program evaluations may also be conducted as a means of program *justification*. Program administrators are frequently asked to justify their resource allocation decisions. Collecting evidence to provide support for the program is another purpose that may be served through a comprehensive program evaluation effort. In some cases, specific evaluation activities may be directed primarily toward program justification issues.

Another purpose of adult program evaluation is that of *budget accountability*. Resource limitations require that program administrators make allocation decisions among a number of program alternatives. Programs that have a demonstrated track record of proper fiscal management and are able to generate surplus revenues (i.e., profits) will generally be viewed more favorably than other programs. Budget reviews may also be conducted to identify expenses that may be reduced and/or revenues that may be increased to improve program profitability and accountability from a budgetary standpoint.

Evaluations may also be conducted to *certify* that the program meets professional licensure or certification requirements. Program standards that specify minimum criteria for accreditation will usually involve some degree of external validation to acknowledge that the standards have been met. Self-study guidelines are often utilized by program staff to examine program standards and gather evidence regarding performance relative to each standard. Site visits may then be scheduled that involve external consultants who are essentially charged with validating the self-study evaluation results, but will also develop a separate report and set of recommendations for program improvement.

Decision Making

The second purpose of program evaluation efforts is for improved decision making. Evaluation efforts should be designed to provide information that can be used in making decisions regarding the management of the adult education program. Again, the primary focus is on making adjustments that will result in program improvement.

Decisions regarding *program management* and *resource allocation* may be influenced by information collected as part of the evaluation process. Facility and resource use are management decisions that can be improved with the use of program evaluation information. Resource allocation involves the combination of human, fiscal, and physical resources that are used in conducting the program. Evaluation information should be reviewed periodically to determine if changes are needed in program management and resource allocation.

Program revisions and decisions regarding *program discontinuation* often originate from recommendations resulting from program evaluation efforts. Some persons fear program evaluations from the perspective that their program might be eliminated. They should recognize that a well-conceived program evaluation should not lead to a recommendation of program elimination without sufficient evidence to warrant such a conclusion. Program discontinuation is more likely to result from the unwillingness or inability of adult educators to keep pace with changes taking place in the environmental context. Responsiveness to evaluation recommendations should positively contribute to program relevancy in times of change, and adult educators should take proactive steps to ensure that their programs operate at the "cutting edge."

Program evaluations have also been conducted to *select award winners* and to *recognize programs for outstanding performance*. Selection criteria will often require evaluation activities to collect comparable evidence that can be used to identify deserving recipients.

Additional decisions that may be influenced by the results of an adult program evaluation include *recruitment, achievement,* and *instructional personnel performance appraisals*. Each of these purposes address target areas for program improvement and may be included as part of an overall program evaluation effort.

Regardless of whether the program evaluation is intended for accountability or for decision making, the intended purpose(s) should be clearly stated at the outset. Evaluation information collected for one purpose, may not be applicable when reviewed for a different purpose. For example, evaluating an adult education program to select an award recipient may not yield valid data that would be useful if the purpose of the evaluation was for budget accountability. Although either alternative is a legitimate purpose for a program evaluation, it is important to define the purpose of the evaluation before collecting evaluation information.

Focus of Evaluation

Comprehensive program evaluation efforts can be classified into four focus areas as identified by Stufflebeam, et al. (1971). The four areas were labeled Context, Input, Process, and Product.

Context Evaluation is involved with analyzing the environmental context in which the adult program is or will be conducted. Another term frequently employed when describing the process of Context Evaluation is "environmental scanning." As the term implies, looking outwardly and scanning the landscape to identify the current situation and emerging trends is the primary focus of Context Evaluation. The general overview question to be addressed in Context Evaluation is: "What are the environmental trends and characteristics that influence the program objectives?" This effort may be completed in conjunction with the needs assessment. However, the view is much more global in perspective, whereas the needs assessment is more likely to examine the needs and interests of individuals in the target audience. Context Evaluation is directed toward issues of need, suitability, practicality, importance, applicability, and validity.

Input Evaluation involves an assessment of all resources that were used in conducting the program. Major resources allocation categories that are usually examined as part of an Input Evaluation include funding, time, facilities, personnel, and equipment. The primary criterion for the evaluation of program inputs should be: "Were the inputs used in the program appropriate and sufficient to achieve the stated objectives?" Input Evaluation should consider both the quantity and the quality of the resources used in the program. For example, in the case of technical equipment, the number of machines available for use in the program would be an important consideration in addition to judging whether the equipment was equivalent to that currently used in real world settings. Input Evaluation should be directed toward issues of efficiency, facility utilization, budgetary requirements, resource use, and input quality.

Process Evaluation is focused on an assessment of the elements of design associated with an adult education program. Most of the design elements involve a variety of alternatives or choices. The primary focus of Process Evaluation is: "Was the overall program design appropriate to achieve the stated objectives?" Process elements include registration, scheduling, content, organization, learning activities, methods, instructors, and assessments. These elements are commonly examined in program evaluation efforts. Although written evaluation forms are often used to collect information regarding process elements, many adult participants provide verbal and non-verbal cues pertaining to their perceptions of many of the process design elements. Adult educators should be perceptive of these cues and be prepared to modify program design features as necessary to provide a more satisfactory and successful program.

Product Evaluation involves an assessment of the results of the adult education program. Achievement and performance measures, including completion rates, placement rates, skill attainment levels, satisfaction ratings, graduation rates, and safety records, are examples of evaluative information that may be examined as part of a Product Evaluation. In most cases, the product refers to the adult participants who completed the program and the level of achievement as measured by the changes in their knowledge, attitudes, and skills. Product Evaluation is focused on formulating an answer to the question: "To what degree did the program achieve the stated objectives?" For many persons, the ultimate measure of program success (i.e., the bottom line) are found in the answers to the questions addressed in the Product Evaluation. However, adult educators should recognize that the Context, Input, Process, and Product components are all important factors to consider in a comprehensive adult education program evaluation.

Personnel Involved in Evaluation

Adult educators generally assume the primary responsibility for program evaluation activities. However, advisory committees also have some degree of responsibility for program evaluation in addition to the planning and organizing functions. Adult educators should be involved in identifying the primary purpose(s) of the evaluation and in planning the overall evaluation effort. The primary role of the advisory committee should be to provide input in planning the evaluation by developing criterion questions to address. The adult educator and the advisory committee should work collaboratively to identify information or evidence to collect to answer the criterion questions.

The adult educator will usually assume the primary responsibility for collecting the evaluative information as directed by the advisory committee. The information should also be summarized by the adult educator and reported to the advisory committee for their analysis and interpretation. The advisory committee is charged with the primary responsibility for formulating recommendations for improvement in the adult education program. These recommendations should include a rationale and justification and be based on information collected as part of the evaluation process.

Program evaluation efforts should be completed as objectively as possible. Sometimes, evaluations appear to emphasize the positive aspects and completely ignore any negative aspects associated with the program. In other situations, the exact opposite may be true; there appears to be an overemphasis on the negative points, while the positive points seem to be glossed over. Either tendency of leaning, whether toward the positive or toward the negative perspective, biases the results and the interpretation of the evaluation effort. Accurate and unbiased evaluation information should be collected and interpreted for accountability and decision making purposes.

The ultimate outcome resulting from the program evaluation process is the formulation of recommendations for program improvement. These recommendations should be developed by the advisory committee and communicated to program administrators. Alternative actions by program administrators in response to the recommendations they receive include acceptance, acceptance with modification, or rejection. Regardless of the action taken, the results should be reported back to the advisory committee in a timely manner. This feedback is essential to acknowledge the efforts of the advisory committee and to demonstrate that their recommendations have not "fallen on deaf ears."

Steps in the Evaluation Process

As previously noted, program evaluation is a continuous process in which the conclusions and recommendations resulting from the evaluation provide feedback information for subsequent program improvement. Figure 13 presents a graphic illustration of the overall program evaluation process.

FIGURE 13.

Program Evaluation

Step 1   DEFINE OBJECTIVES

Review the stated goals and objectives of the adult education program and define the purpose(s) of the evaluation effort. Develop a plan for conducting the evaluation and determine which of the major components (i.e., Context, Input, Process, and Product) should be addressed.

Step 2   DEVELOP CRITERION QUESTIONS

Utilize advisory committee input to develop a list of questions that should be answered to fulfill the purpose(s) of the evaluation. The questions should be specific and stated in measurable terms. It is preferable to have a greater number of questions stated in specific terms, than to have fewer questions stated in general terms.

Step 3   IDENTIFY ACCEPTABLE EVIDENCE

Each criterion question should be analyzed to determine the information needed to appropriately and completely answer the question. Strategies should be developed to plan for the collection and summarization of the evaluative information.

Step 4   ANALYZE AND INTERPRET THE INFORMATION

Information collected to answer the criterion questions should be presented to the advisory committee for analysis and interpretation. The evaluation information should be reviewed with regard to each program objective as well as the overall program goals.

Step 5   FORMULATE RECOMMENDATIONS

The advisory committee should be charged with developing recommendations based on the evaluation information. The recommendations should focus on areas targeted for improvement in the program and include a statement of rationale and justification.

Step 6 **REPORT TO DECISION MAKERS**

The results of the program evaluation should be reported to the decision makers (i.e., program administrators, board of directors, director of extension, Vice President for Human Services, etc.). The report should outline the original program objectives and the purpose(s) of the evaluation. In addition, the report should describe the major findings and observations resulting from the evaluation effort and transmit the recommendations developed by the advisory committee. The report should conclude by asking the decision-making authorities to take action on the recommendations and provide feedback to the advisory committee pertaining to the action taken.

The actions taken in response to recommendations offered by the advisory committee will generally be directed back toward the program. In response, the adult educator in conjunction with the advisory committee should be prepared to take action to improve the program based on the response from the decision makers.

Summary

Program evaluation should be viewed as a continuous loop that involves the collection of information needed to answer criterion questions. The evaluation effort should involve the joint efforts of the adult educator and the advisory committee. The primary purposes of a program evaluation is to address issues of program accountability and to provide information to guide and direct decision making for program improvement.

CHAPTER
11

PROMOTING
ADULT PROGRAMS

Many adult education programs have failed because of poor promotion — people just did not know about the program or did not realize how good is was. One characteristic that frequently differentiates adult education programs from other forms of education is voluntary attendance. Adults usually have the ability to "vote with their feet" (Imel, 1994) thereby enabling them to make a deliberate choice of whether or not to participate in the program.

However, before adults can be expected to make an intelligent decision regarding their participation, they must first be informed about the costs and benefits of the program. Therefore, promotional efforts are very important to the success of adult education programs and should be planned well in advance to allow sufficient time for the desired message to reach the target audience.

Purposes of Promotion

The most obvious need for promoting an adult education program is to encourage enrollment and participation. Many adult education programs are operated on a cost-recovery basis and must have a minimum threshold of fee-paying registrants to recoup the costs of planning and conducting the program. Adult programs that do not generate sufficient enrollment to cover the costs will usually be canceled unless another source of revenue can be arranged. Adult educators should recognize that adult education programs are scrutinized more closely with regard to income and expenses than most elementary and secondary school programs. This reality is based on the philosophical view that tax-supported public education in this country is provided free of charge up through grade twelve. Educational programs beyond that level (i.e., higher education and adult education) must be financially supported through other sources (although many local school districts subsidize overhead costs for electricity, room rental, audio-visual equipment, etc.). Further discussion of the funding of adult education programs will be presented in Chapter 12.

Promotional efforts should be planned that communicate the anticipated benefits of participation. Individual members of the target audience are usually "tuned into radio station WII-FM"—that is, they are interested in "What's In It For Me?" Therefore, adult educators should recognize the need to promote the program from the perspective of individual benefits for participants. Focusing on the *wants* of the target audience (see Chapter 7, Needs Assessment) will help ensure that the promotional effort "strikes a chord" with individuals who will subconsciously say to themselves "This is a program in which I want to participate!" Although more global benefits may also be identified and promoted, individuals who are free to choose whether or not to participate will be more likely to base their decision on personally anticipated benefits.

Adult education programs that involve mandatory participation are faced with a uniquely different challenge in promotion. Programs that require attendance should communicate the manner or degree to which participation in the program meets CEU (Continuing Education Unit) or licensure requirements in addition to individual and personal benefits. The greatest barrier that must be overcome with adult education programs that mandate attendance is the experience of participants who failed to recognize individual benefits from previous program participation. Overcoming prior experience is a significant hurdle for many adult educators.

Adult education programs frequently receive significant benefits from endorsements by professional associations, labor unions, and employers. Such endorsements constitute a "seal of approval" that provides external validation of the value of the anticipated benefits. Securing the endorsement of one or more support groups presents a unique challenge for adult educators. In addition to highlighting the personal and individual benefits, promotional efforts among potential sponsors will need to address more global benefits as well. Holistic benefits are important factors to communicate to support groups to secure an endorsement for an adult education program. However, such benefits may be more difficult to define and to measure. Support from other groups is a significant benefit in promoting adult education programs. Such support is often viewed more objectively by the target audience, whereas promotion by the adult educator (provider) may be viewed with some degree of skepticism as being potentially biased.

Promotional efforts may also be directed toward publicizing adult education programs to a broader community beyond the intended target audience of potential participants. Elevating community awareness of adult education opportunities can produce unexpected benefits. Adult education programs may uncover previously untapped community resources to support current or new program offerings. Conversely, the community members may increase their awareness of the possibilities for planning and conducting adult education programs in subject areas that had not been considered previously. The potential of unanticipated mutual benefits is significant and should be recognized by adult educators. They should constantly be aware of opportunities to promote their programs within the broader community. Frequently, these opportunities arise under unusual circumstances and are unplanned. Adult educators should take full advantage of such situations. Adult educators need to be prepared to "toot their own horn" for their programs.

Steps in Promotion

Promoting adult education may range from a loosely structured "word-of-mouth" approach to a more formalized and highly structured sequence of activities. Regardless of the approach employed, adult educators should follow some general guidelines as they plan and implement promotional activities.

Step 1   SET AN ENROLLMENT GOAL

The first step in the promotional process is to establish a realistic goal for the number of participants. Several factors should be considered in setting the enrollment goal. Most importantly, "What is the overall objective of the program?" Programs directed toward developing the technical skills of individual participants should be limited to fewer participants than programs designed to introduce new concepts (without an individual skill development component). The objective(s) of the program is (are) the primary determining factor that influences the number of participants that can be accommodated.

Other factors that should be considered when establishing an enrollment goal include:

1. Size of the target audience

2. Size of the facility (room, seating, work stations, equipment, etc.)

3. Number of resource persons (lab assistants, technicians, etc.)

4. Cost of program

5. Location of program (geographic distribution of target audience, travel distance, etc.)

6. Duration (length of program)

Each factor should be a consideration in establishing a realistic enrollment goal. However, the overriding factor should be the program objectives and anticipated outcomes for the participants.

Step 2   DEFINE THE TARGET AUDIENCE

The second step in the promotion process involves defining the target audience. Specifying, as distinctly as possible, the characteristics of the clientele group the adult education program is designed to serve is extremely important. Promotion is primarily a communications process and to communicate the message effectively you must "know your audience."

One primary consideration is whether the target audience is relatively homogenous or heterogeneous. This factor will be influenced to a large degree by the program topic. The more similar the characteristics of the target population (i.e., homogeneous) the easier it will be to focus the promotional effort. Characteristics that affect the variability of the target audience include geographic distribution, education level, socio-economic status, marital status, age, career status, and organizational affiliation to name only a few. The variability within each of these characteristics among the target audience will affect the ability of the adult educator to promote participation in the adult education program.

The use of communication tools, such as radio, television, newspapers, newsletters, etc., is influenced by the geographic distribution of the target audience. The general rule should be to use the most individualized and personalized communication media possible in the promotional program. At one extreme, word-of-mouth communication between known and respected individuals is one of the best forms of promotion. The opposite extreme would involve mass communication tools, including radio, television, or box holder mail, to blanket the promotional message to a broad range of potential participants.

Regardless of the communication tools used, there tends to be a trade-off between their cost and their effectiveness in stimulating participation. Personal visits, letters, and phone calls are generally the most effective methods for promoting participation in an adult education program. However, they are also the most expensive in terms of the number of persons contacted. Mass media tools, such as radio, television, and newspaper advertising, may have the lowest cost per person contacted. However, the effectiveness (as measured by percent of target audience participation) is relatively low.

Promoting adult education should usually involve a combination of the common promotional tools to reach the enrollment goal. Also, resource limitations of time and money may preclude the use of certain promotional tools although they may be more effective.

Accessibility to the target audience must also be considered in developing a promotional plan. Having access to a mailing or telephone list of individuals in the target audience makes it physically possible to personally contact each individual to encourage his/her participation. Therefore, the more directed the promotional effort, the more successful. Lacking direct access to names, addresses, and telephone numbers of individuals in the target audience requires that more indirect communication tools be utilized to convey the promotional messages. The use of mass media tools results in a shotgun approach to entice participation from individuals who receive the promotional message. The use of mass communication results in a "hit or miss" approach to promoting enrollment in adult education programs. However, in the absence of direct access to individuals, mass communication tools may be the only feasible alternative available for use in promoting adult education programs.

Step 3   **PLAN THE PROMOTION STRATEGY**

The third step in the process of promoting an adult education program is planning. Often, the promotional activities may be limited to a single tool or approach, involving a program description in an adult education course schedule, a newspaper advertisement, or a tri-fold brochure. In some instances, the responsibility for the promotional efforts may be deferred to a program administrator or another individual, who may be promoting a variety of program options. As a result, each program is promoted in a rather generic fashion. To maximize the effectiveness of the promotional effort, the adult educator should be involved in the promotional process. Involvement of an advisory committee would capitalize on "group thinking" to plan the most effective promotional strategy. Dividing responsibilities among members of the advisory committee produces an increased sense of commitment and ownership in the program.

The promotional plan should be viewed as a comprehensive process rather than a single activity to be completed before the program begins to stimulate interest and enrollment. Communication tools and promotional opportunities should be planned before, during, and after the program. Prior to the program, the promotional effort should be directed toward a target audience of potential participants to encourage enrollment. During program operation, promotional activities might equate to progress reports to inform administrators and the general public of program activities. After the conclusion of the program, promotional efforts should focus on communicating the results of the program for the participants as well as the general public.

For each stage of promotion (i.e., before, during, and after), the message to be communicated and the target audience may differ slightly. However, each of the stages and messages provides opportunities for public relations benefits for the adult education program.

Step 4   **PREPARING AND DISTRIBUTING PROMOTIONAL MATERIALS**

The fourth step in promoting an adult education program is the development and distribution of promotional materials. Again, the rule is the more targeted the message, the more effective. The characteristics of commonly used promotional tools are presented in Appendix H.

Printed materials should be prepared to take into account the reading level of the target audience by conveying the intended message in a clear manner. The use of color, graphic illustrations, charts, and pictures can be helpful in communicating key concepts as part of the promotional message.

Audio or audio-visual messages can also be used to promote adult education programs. These media should be carefully developed to ensure the appropriate message is being conveyed in a clear, concise, and consistent manner. Often, mass communication tools can be used to generate awareness or stimulate interest in an adult education program, although more specific information regarding registration may be provided through another contact source. One-way tools of mass communication should always contain a name, phone number, address, e-mail, or Internet home page URL of who to contact for additional information about the adult education program or related information.

Printed materials in the form of flyers, brochures, and form letters are frequently used to promote enrollment in adult education programs. These materials should be designed specifically (or modified) for each target audience. The primary message for the printed material should relate to personal and individual benefits of participation. In addition, to create a sense of urgency for enrollment, some promotional materials suggest that enrollment is limited, early bird discounts apply, and/or there is an enrollment deadline.

Step 5   EVALUATE THE RESULTS

The most important step in the process of promoting adult education programs is to evaluate the results of your efforts. The primary questions to be addressed in the evaluation are "How did the participants learn about the program?" and "What factors were most influential in their decision to participate?" Answering these questions will identify promotional activities that were successful and those that were not successful in generating enrollment. Often, the information needed to analyze the promotional effort is best collected in conjunction with the enrollment or registration process. Registration forms can be designed to solicit quick responses from participants regarding factors that contributed to their decision to enroll in the program.

Effective use of promotional tools requires a planned effort that takes into account the needs and interests of the target audience. Using

FIGURE 14.

Adult Education Registration Form

Adult Education Registration Form

Name _____   Day Phone _____
Address _____   Evening Phone _____
City _____ State _____ Zip _____

Payment Method   Check   Credit Card   No. _____   Exp. Date _____

Program name _____   Amount $ _____

Signature _____

How did you learn about this adult education program? (Check all that apply)

_____ Brochure   _____ Newspaper   _____ Flyer   _____ Friend   _____ Employer   _____ Other

What is the primary factor that prompted you to enroll in this program?

_____
_____

For additional information contact: Make checks payable and mail to:
    Adult Education Coordinator   City Public Schools
    City Public Schools   Adult Education Program
    (555) 379-5831   Phone   Box 150
    (555) 378-5612   FAX   City, ST 99999-9999
    adulted@city.school.edu   e-mail

---

a variety of promotional tools is generally more efficient than promoting adult education with a single tool. Company or school letterhead should be used whenever possible to generate and maintain institutional recognition. Name recognition is a significant advantage in subsequent promotional efforts.

Radio and television materials should be typewritten and double-spaced. Copies of all promotional materials should be maintained on file so they can be modified for later use if they proved to be successful. In addition, materials may be examined at a later date to identify potential problems if they were not found to be effective in promoting participation in the adult education program. It is very important to analyze and evaluate materials that were successful as well as materials that were not successful.

Action photographs should be taken and maintained on file to be used in subsequent adult education promotional efforts. The adage "One picture is worth 1000 words" is especially true in the promotion of adult education programs. Adults who can relate to pictures of other adults in promotional materials may be able to overcome psychological barriers that may have otherwise inhibited their participation. Free publicity may be garnered by alerting the news media to a newsworthy story. Planning a media event in conjunction with an adult education program is another avenue that can be pursued. However, take care to ensure that the intended message is what actually gets reported.

In addition to evaluating the promotional tools that assist in generating enrollment in an adult education program, it is equally important to analyze factors that inhibit potential enrollees from participating. Surveys of non-participants are helpful in identifying barriers to participation, although the surveys may be very difficult to complete in an efficient manner. Most notably, it is often difficult to identify individuals in the target audience who chose not to participate. One alternative is to contact individuals who inquired about the program, but did not enroll. Although this practice may be used to identify barriers to participation within one subset of the target population, the barriers may be different for individuals who had an interest in the topic, but did not inquire about the program.

The inability to access individuals who did not inquire about the adult education program is a limiting factor in planning subsequent promotional efforts. Knowing why individuals did not enroll is equally important to knowing why others chose to enroll. Information from both perspectives would certainly improve the promotional efforts for adult education programs in the future.

Summary

Promoting adult education is an important function that differs significantly from most elementary and secondary education programs. Recognizing that participation is a voluntary decision made by individuals suggests that promotional efforts should be directed toward individual needs and interests. Adult educators who are attuned to the needs and interests of the target population can use that information to help guide the promotional effort.

Promoting adult education should be viewed as a continuous process and incorporate a variety of communication tools. The process should begin with establishing an enrollment goal, defining the clientele, planning the promotional strategy, developing materials, and evaluating the results of the promotional effort. Successful adult educators recognize the need for promotion and capitalize on opportunities to promote their programs as the opportunities become available.

CHAPTER 12

FUNDING ADULT PROGRAMS

One primary feature that differentiates most adult education programs from other educational programs involves the manner in which they are funded. Elementary and secondary educational programs in public schools receive their primary funding support from state tax revenues in the form of appropriations (or distribution formulas) from state legislatures and administered by state departments of education. Although the system of funding differs somewhat from state to state, as a general rule the funding to support public elementary and secondary education is provided by state tax revenues, with some additional support from the federal government.

Funding for higher education programs, such as community colleges and other public colleges and universities, is obtained from a variety of sources. In addition to some state legislative appropriations, these institutions may receive revenues from student tuition and fees, federal appropriations, external grants, and private contributions. The mix of revenue sources among higher education institutions also varies widely. However, the general trend is a decrease in the proportion of total revenue provided through continuing state appropriations. This trend has contributed to the need to seek other revenue sources (e.g., grants, gifts, etc.) to maintain or increase funding levels.

Funding for adult education programs is unique in that state and federal tax revenues tend to be limited to adult education programs that are remedial (i.e., Adult Basic Education [ABE], General Education Development [GED]) or employment oriented (i.e., English as a Second Language [ESL], displaced homemaker, Job Training Partnership [JTPA], etc.). Therefore, the primary source of funding for most adult education programs in the United States is from participant fees or sponsorships.

Recognizing that most adult education programs are operated on a cost-recovery basis places a degree of fiscal accountability on adult educators. Although most adult education programs are administered through an institution (e.g., public school, community college, university, labor union, etc.), there is an implicit assumption that each program will secure sufficient fiscal resources to pay all expenses incurred and, in many cases, generate income above program costs. In fact, many educational administrators view adult education programs as a "cash cow" or a profit-making center that provides budget flexibility for the overall educational enterprise.

The process of funding adult education inevitably involves the development of a budget. Budgets should be viewed as a tool in the planning process that translates the resources required to fulfill program objectives into quantitative terms. Program budgets are essentially estimates of the physical and financial resources needed to implement the program. The primary difficulty in preparing such budgets is due to the number of "unknowns." Frequently adult educators do not know how many participants to expect. Nor, do they know the cost of publicity, the cost of travel, or the expense associated with audio-visual equipment. Although the ability to develop a budget increases with experience, first-time adult educators experience much frustration over the challenge of developing an adult education program budget. The purpose of this chapter is to provide some basic guidelines and factors to consider in developing budgets for adult education programs.

Developing a Budget

Developing a budget for an adult education program is an important aspect of the overall planning process. The program budget involves an estimate of program revenues from all sources and the expected expenditures involved in conducting the program. Therefore, from a simplistic point of view, a program budget is organized as follows:

Program Revenue (income)
−   Program Costs (expenses)

Profit (Loss)

Preparing a budget should begin with identifying sources and amounts of revenues and costs. However, with most adult education programs, there are very few items of revenue or costs that are "known," and the budget development process shifts to one of "estimation."

Expense Budgeting

One strategy is to focus on estimating program expenses. Adult education program budgets may be organized into five major expense categories (see Broomall, J. K. & Fisher, R. B., 1995), as follows:

Promotion and advertising — These expenses include all costs associated with informing and encouraging participation in the adult education program. Although promotional efforts are addressed in greater detail in Chapter 11, the costs of all promotional activities need to be incorporated into the program budget. Mass media, such as radio and television advertising, may be quite expensive. However, the size of the target audience and the potential for significant participation may justify using high-cost media in the promotional effort. On the other hand, personal letters may be more effective with a smaller, well-defined audience in which case the cost of publishing, postage, and materials should be included in this budget category. Telephone solicitation, travel for personal promotion, brochure development and printing costs, photography, and videotape production and editing are examples of promotional activities that should be considered when developing the Promotion and Advertising budget category. Promotional activities typically comprise 10 to 15 percent of the total direct costs.

Faculty — Personnel costs for individuals involved in planning, conducting, and evaluating the adult education program may be included in the faculty budget category. Both salary and fringe benefits should be included for each faculty member involved in the program. Often, part-time faculty are budgeted at a fixed rate per instructional contact hour. These rates may vary widely between the minimum wage rate to over $30.00 per hour. In some instances, adult education program resource persons will range from "no cost" to an honorarium of several hundreds or thousands of dollars plus expenses. Obviously, the cost of high-priced speakers should be agreed to in a written contract to ensure that revenues are sufficient to justify those expenses. Corporate or industry sponsorships are often secured to pay all, or a portion, of the costs of "keynote" speakers at major conventions, or to secure speakers who command substantial fees for smaller-scale adult education activities. Securing external funds for speakers via sponsorship will usually involve some degree of commercial, or corporate, advertising or endorsement in return for the sponsorship. This should produce a win:win situation in which everyone derives some degree of benefit. In some instances, guest speakers may decline an honorarium (e.g., government employees). However, in such cases, reimbursement for their travel expenses may be justified.

Local institutions involved in administering adult education programs will generally have some administrative guidelines or policies that address faculty salary and benefits. These guidelines should be reviewed to determine the degree of flexibility that exists in calculating budget estimates in the faculty category.

Instructional support — Estimating the cost of facilities, equipment, resources, supplies, and materials needed to conduct an adult education program can be an imprecise activity. While planning adult education programs, it may be difficult to anticipate the full range of expense items that may be encompassed within the instructional support category. Programs conducted on the site of educational institutions may have free access to the use of audio-visual equipment and computer laboratory facilities. However, the same program conducted at a commercial site (i.e., convention hotel, etc.) may incur several thousands of dollars of audio-visual expenses. Photo reproduction costs, transparencies, slides, binding proceedings, and registration materials are but a few of the items that should be considered when estimating instructional support costs. Again, institutional support may have a significant effect on whether or not these items may be provided without a direct charge to the program budget or whether every anticipated cost will need to be included. Again, institutional guidelines and policies should be considered in the formulation of this budget category.

Travel — Expenses for travel may include faculty, resource persons, and, in some cases, participants (e.g., field trips, etc.). Mileage rates and per diem lodging and meal rates may be limited by institutional travel policies. It is recommended that budgets be developed based on the maximum allowable rate to ensure sufficient funds will be available to cover all necessary and allowable expenses. Furthermore, receipts for all travel, lodging, and meal expenses should be collected and maintained on file to document expenses attributed to the adult education program budget. Institutional travel reimbursement forms should be provided to resource persons before they leave the program site so they can submit their expenses and receipts for reimbursement in a timely manner after they return home. Honoraria checks may be prepared in advance and hand-delivered to speakers upon completion of the program to avoid a prolonged delay in processing the paperwork for payment after the meeting.

Overhead or indirect — This category is comprised of all expenses necessary to administer, manage, and maintain the program, but cannot be explicitly attributed to an individual program. Such costs may include utilities (electricity, heat, air conditioning, water, sewer, and telephone), building and facility maintenance, fiscal management, accounting, purchasing, payroll, registration, and other functions important to the success of the program. These functions represent a true cost to the institution but cannot be directly attributed to individual programs that benefit from the services (thus the term "indirect" costs).

Institutional guidelines and policies usually specify how adult education programs should budget for indirect costs. Such calculation methods include (a) a percent of total direct costs (or some derivative of total direct costs), (b) a flat fee per participant, (c) a flat fee per program, or (d) some combination of the three. The indirect costs included in the program budget are usually transferred to the institution's general operating fund account and thereby made available to pay for expenses associated with administrative services and utilities.

Revenue Budgeting

After the expenses for the adult education program have been listed, attention should be directed toward estimating sources and amounts of revenue to support the program. Participant fees are frequently the sole source of revenue for adult education programs. Therefore, the total expenses for the program should be divided equally among the total number of participants. To calculate the fee per participant, the adult educator needs to specify a minimum number of participants needed to operate the program. This threshold number should be somewhat conservative yet realistic with regard to the potential for the actual number of participants. Establishing the minimum number of participants is one of the most arbitrary and subjective decisions that adult educators make in the planning process. However, the decision to establish a minimum enrollment may significantly affect participant fees and, ultimately, whether the program may have to be canceled due to insufficient enrollment. The calculations presented below illustrate the effect on registration fees when the minimum number of participants is changed between 20 and 30.

		Minimum of 20	Minimum of 30
	Total Expenses	$2,400	$2,400
÷	Minimum Number of Participants	÷ 20	÷ 30
	Registration Fee per Participant	$ 120	$ 80

Although participant fees are often the sole source of revenue for adult education programs, other sources may be tapped to fund all or a portion of the expenses. Customized training programs designed for employees of an individual company often are funded 100 percent by the company with no charge to the individual participants. Government sponsored programs, safety programs, and public information programs are other examples that may receive 100 percent funding from a source other than the participants. The latter examples have the added challenge of promoting the quality and value of the programs when they are free to participants who generally hold the view that "You get what you pay for." Some adult educators suggest that a nominal fee should be charged to prevent uninterested participation of persons attending (just because it is free) and to acknowledge that the program has inherent value.

Some adult programs receive funding support from a combination of sources. Local businesses and industry groups use sponsorships to promote the value of the program while simultaneously receiving recognition (in the form of public relations) for their support. The revenue generated from partial sponsorships should be subtracted from the total expenses before dividing the remaining expenses by the number of participants in calculating registration fees.

Example:

Total Expenses	$2,400
Sponsorship	− 600
Expenses After Sponsorship	$1,800
Minimum Number of Participants	÷ 30
Registration Fee per Participant	$ 60

The calculations above were computed under the assumption that revenues at some minimum threshold of participation should match the total expenses for the adult education program. Formulating a program budget in which registration fees (in addition to other revenue sources) establishes the minimum participation level as the "break-even" point. In some instances, it may be desirable to establish registration fees at a somewhat higher level to generate surplus revenues or profit. Participation rates above the minimum threshold level would also produce more revenue and increase the amount of profit resulting from the program.

Some institutions seek to generate profit from adult education program offerings. In such cases, development of the program budget may include a flat amount of profit the program should generate or add a flat amount (i.e., surcharge) to the registration fee of each participant. Either strategy can be effective as a means of profit generation and provides a certain degree of "cushion" in the budget in the event that expenses were underestimated for some reason. Establishing the final registration fee for participants is crucial to the overall success of the program. If the target audience views the registration fee as too high, it may be a barrier to participation. However, in some instances, the target audience may view a relatively low registration fee as a negative indicator of the program's quality.

Knowledge of the target audience is helpful in establishing an appropriate registration fee. "Early bird" registration discounts and late registration surcharges can also be used to encourage early registration. In addition, recognizing if the fee will likely be paid by an individual or whether a third party (e.g., employer, scholarship, etc.) will pay the fees for participants, may have some influence on the final determination of the registration fee.

As a general rule, in the budget-building process it is best to estimate expenses on the high side and revenues on the conservative side.

This practice will allow for a degree of error in estimation without resulting in a net loss.

A sample program budget worksheet follows. This form provides one example of how an adult education program budget can be developed to account for the six major categories of expense items, revenues from sponsorship, establishing a minimum threshold of participation, and the calculation of a break-even rate for registration fees.

FIGURE 15.

Sample Budget Worksheet

Adult Education Program Budget					
Expenses			**Revenue**		
	Budget	Actual		Budget	Actual
Promotion Advertising Brochures Postage Travel			**Donations:**		
Faculty Salaries Benefits Honoraria			**Sponsorships:**		
Facilities & Meals Room Rental A-V Rental Signs/Banners Meals/Breaks Equipment			**= Total Donations and Sponsorships**		
Instructional Support Photocopying Supplies Resource Materials			**Total Expenses** (from left column)		
Travel Buses Lodging Per Diem			**– Total Donations and Sponsorships** (from above)		
= Total Direct Costs			**= Total Net Expenses**		
+ Indirect Costs			**+ Minimum Participation**		
			= Expenses per Participant		
= Total Expenses			**+ Profit Margin/Participant**		
			= Registration Fee		
			Profit (loss)		

Summary

Developing a budget for an adult education program is a complex and subjective process. The budget should be viewed as a "best guess" estimate and actual expenses, revenues, and profit (or loss) should be compared to the budgeted figures to improve the budget-building process for future programs. Revenue and expense records should be carefully recorded and maintained to provide for fiscal accountability.

CHAPTER 13

SELECTING ADULT EDUCATORS

Selecting appropriate personnel to provide leadership for adult education programs is critical. The success of any educational activity is largely dependent upon the efforts of the educator. However, since adult education generally involves a more self-directed and participant-oriented approach, the skills needed to be successful may differ slightly from educators in primary and secondary schools, colleges, and universities. This chapter is devoted to an examination of the skills, abilities, and attitudes essential for adult educators to effectively facilitate learning with adults.

Knox (1980) suggested that adult educators should possess knowledge and skills in three foundational areas. These areas are graphically illustrated in Figure 16. Each of the three areas is an important prereq-

FIGURE 16.

Foundational Knowledge and Skills of Adult Educators

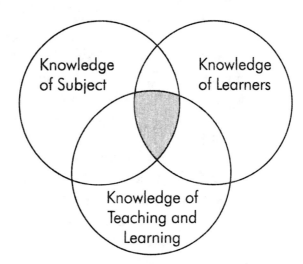

133

uisite for educators to be successful in adult education. However, even more important is that truly effective adult educators recognize how the three domains overlap and that the area of convergence (i.e., shaded area) indicates ideal opportunities for adult learning.

When viewed at the extremes, some individuals may be characterized with circles representing the three domains with no area of convergence. Such individuals cannot envision a relationship between the content and the learners, the content and teaching/learning processes, and the learners and the teaching/learning process.

The opposite extreme would involve individuals who recognize a significant area of overlap between the content, the learners, and the teaching/learning process. Such individuals can envision numerous opportunities to promote the learning process with adults utilizing a variety of teaching/learning strategies adapted to the subject matter content.

Obviously, individuals in the latter group would be more desirable candidates for adult educator roles. These individuals would be expected to view their role as facilitators of adult learning, recognizing that each of the three domains significantly influence the learning process.

Although the three domains outlined are important, as well as the relationship (i.e., overlap) between the three areas, it is the "balance" among the three areas that is the most significant concern. Frequently, adult educators are employed because they are highly skilled and respected as content or technical experts. This expertise is very important to gain and maintain the respect of adult learners. Credibility among adults is very important and directly influenced by the audiences' perception of the adult educator's content expertise or experience. However, there are numerous examples of situations in which an adult educator with excellent technical content credentials was ineffective in meeting the needs of adult learners in the audience. Often, the criticism is expressed as "That teacher was lousy!" It should be recognized that content expertise is important, at least to some baseline level; however, content expertise alone does not equate to success.

Successful adult educators should be recognized experts in their content area and have some degree of familiarity with the target audience, including their needs and interests. Adult educators should also possess some baseline level of knowledge and skills of the teaching/learning processes. To select effective adult educators, administrators, or program managers should review applicants with an eye toward assessing the degree to which they meet at least baseline requirements in each category. Furthermore, job descriptions should be structured to ensure that applicants are informed that knowledge and skills in each of the three domains are requirements for the job. A brief example of suggested "required qualifications" for effective adult educators is provided in Figure 17.

FIGURE 17.

**Position
Announcement**

Position: Adult Educator
Required Qualifications:
1   Content expertise and experience.
2   Knowledge of adult teaching/learning processes.
3   Recognition and appreciation of the diversity of needs and interests of adult learners.
Preferred Qualifications:

_____

_____

_____

Personal Qualities

Adult educators are in a unique position (by virtue of the leadership role they have in an educational environment) to contribute to the educational, personal, social, and economic development of the adult participants. This creates a situation in which the adult educator has an awesome responsibility. Some adult educators intentionally shy away from fulfilling that responsibility, while others fully embrace that role recognizing the potential contribution to the success of each individual adult participant. The differentiating quality appears to be based on the philosophical perspectives of the adult educators regarding their role in the teaching/learning process. Some adult educators view their primary responsibility to be the transmission of content knowledge or skills within and among adult learners. Other adult educators perceive their role to extend beyond transmitting knowledge and skills to include how that information may be utilized or applied to develop individual learners either personally, socially, and/or economically. The latter group of adult educators tend to gain a greater sense of self-satisfaction from the adult education program. This self-satisfaction is a motivating force and prompts many adult educators to continue their involvement in subsequent programs. Although many adult educators are hired on a part-time or as-needed basis, the economic incentive is often out-weighed by the sense of satisfaction received from making a positive contribution to the lives of others. Many adult educators acknowledge it was the monetary income that initially attracted them to become an adult educator. However, it was the sense of self-satisfaction that motivated them to continue their involvement.

Additional characteristics of effective adult educators include:

Organized — Adult educators need to have the ability to organize materials and learning activities to ensure that programs run smoothly. Disruptions caused by a lack of organization interfere with the learning process and create an unnecessary barrier to learning. Organized and efficient learning activities are the expected norm but may not be outwardly acknowledged by the participants. However, disorganized activities are inefficient and may cause adults to discontinue participation if they feel they are wasting their time.

Flexible — Adult learners are individuals who have unique background knowledge, experiences, and environmental conditions. Adult educators need to recognize these individual differences and avoid imposing rigid requirements on adult learners unnecessarily. Most adult learners will strive to fulfill expectations but also appreciate understanding adult educators who make allowances when unique problems or situations arise. Such accommodations should be made (or not made) depending upon whether the "process" or the "product" is the primary desired outcome. Flexibility should be provided in instances when quality standards of the desired outcomes will not be compromised.

Flexibility also relates to rigidity in thinking. Many adult educators report that interacting with adult learners has contributed to their own learning by causing them to view the content from a different perspective. Certain subject areas, such as professional certification or licensure (e.g., emergency medical technician, welding, pilot training, etc.), may require strict adherence to accepted guidelines and, therefore, a certain rigidity in one's thinking. However, in other content areas, differing viewpoints should be encouraged and openly discussed.

Enthusiastic — Adult educators should be passionate about their content area. An attitude of excitement among adult educators for their subject is transmitted as a contagious enthusiasm among adult learners. On the other hand, adult educators who appear to be bored with the content they are teaching also have a tendency to transmit that message to adults. Adult educators need to project genuine enthusiasm for their subject. False or insincere efforts to appear enthusiastic will generally fail over time. However, presenting a realistic attitude of excitement will generally produce positive results in the adult teaching/learning process.

Dedicated/committed — Many adult educators have full-time jobs in addition to their adult educator responsibilities. In addition, they have time commitments and environmental pressures that are similar to other adults. However, accepting the role as an adult educator should require a certain level of dedication or commitment. This characteristic implies that the adult educator leaves personal problems at the doorstep and projects an image that the adult learners are the highest priority during the program. Adult educators who rush in immediately before the beginning of the program or who disappear before the adult participants have left the room, are subliminally sending the message that they are too busy to attend to the needs of the participants. Adult educators should arrive before and leave after the participants (if at all possible) to project a sense of dedication and commitment. Often, the informal interactions between adult educators and participants before and after a formal program are the most beneficial and informative.

Perceptive/empathetic — One of the primary responsibilities of an adult educator is to develop a climate (physically and psychologically) that is conducive to learning. Effective adult educators need to be cognizant of verbal and non-verbal cues provided by participants. Restless behavior in the audience may indicate the need to take a short break or that there is some disagreement or misunderstanding among the audience. Adult educators should develop the ability to ask thoughtful and probing questions, but avoid intimidating the audience in the process.

Some adults have a very narrow range of tolerance for heat, light, air movement, noise, and other factors that may potentially interfere with their learning. Some adults may be quite outspoken regarding a discomfort they are experiencing. Other adults will politely tolerate the situation to avoid the appearance of creating a nuisance or "causing trouble" for the adult educator.

Providing feedback to adult participants is another important role of adult educators. Communicating clearly and tactfully about progress will help to reinforce satisfactory performance and elevate the self-confidence of participants. Adult educators should be cognizant of the fact that many adults will attempt to mask their lack of confidence with various patterns of behavior as a defensive mechanism. Developing and maintaining good interpersonal skills is vital to the success of an adult educator (Galbraith, 1989).

Effective as a facilitator — One of the most important qualities of an effective adult educator falls outside the realm of content knowledge or skill. The primary role of the adult educator should be to serve as a facilitator of adult learning. That role is best fulfilled when adult educators view themselves as a coach, mentor, or supporter rather than a purveyor of knowledge. This characteristic is governed more by the adult educator's attitude and philosophy than by subject matter expertise. Frequently, adult educators who are recognized for their technical competence must learn to re-focus their priorities from a content-orientation to a learner-orientation. For many adult educators, this requires a paradigm shift from the content to the learner as the priority. In addition, many business-oriented adult educators have been trained to focus on a bottom-line product, whereas many adult education programs should concentrate on *process* outcomes. Therefore, adult educators should focus their primary efforts on teaching adult participants "how to think" rather than "what to think."

Staff Development for Adult Educators

Ideally, adult education personnel would have expertise in the content, be knowledgeable of adult teaching/learning processes, and be able to recognize the needs and interests of their audience. However, in reality, individuals who meet all three qualifications are rare. Full-time adult educators are more likely to have these strengths, which is why they were attracted to adult education as a career field.

Part-time instructors account for 80 percent of all adult educators (Bankirer, 1995) and, generally, possess a different set of strengths as compared to full-time adult educators. Part-time adult educators are usually employed in an occupational area related to their subject matter content. These instructors' greatest strength lies in their professional competence and practical application of their content knowledge (Bankirer, 1995). However, they are frequently less knowledgeable and experienced in the teaching/learning processes and the personal needs and interests of the target audience.

Recognizing the relative strengths and weaknesses of part-time adult educators, there is a need for pedagogical assistance to enhance success in adult education programs. Many institutions, which employ part-time adult educators, offer orientation sessions to acquaint new instructors with their role as the facilitator of adult learning. Such programs need to be well planned and well conducted to model effective teaching/learning activities and to avoid offending individuals who may hold the misguided opinion that "anyone can teach."

Adult educator orientation sessions should be comprehensive in nature by creating an awareness of institutional and community resources that may be utilized in the adult education program. Institutional policies that are important to the adult educator and program participants should be reviewed. Adult education policies and procedures handbooks are frequently distributed as a resource for part-time adult educators. The handbooks are useful when problems arise related to parking, registration, fee refunds, food and drink availability, photocopying, library and computer accessibility, etc.

Faculty development opportunities should be provided (possibly even required for first-timers) for guidance in the process of planning, conducting, and evaluating adult education. For many adult educators, their only experience in education was from the perspective of a student. Although viewing education from a student perspective is beneficial, it is certainly not sufficient preparation to be an effective adult educator. Most notably, the pedagogical practices of taking daily attendance, assigning seats, conducting weekly quizzes, dealing with homework problems, and giving multiple-choice exams may not be appropriate. However, if those are the only educational experiences known to the novice adult educator, he/she most likely will be inclined to mimic those practices in teaching adults. Orientation programs for new (first-time) adult educators should be provided to avoid such pitfalls.

Using Resource Persons

Resource persons are frequently used in educational programs as guest speakers or panel members. Such persons should be utilized when it is desirable to have someone with special expertise that the regular instructor does not possess. Resource persons are particularly helpful in providing up-to-date information, technology, or practical applications of the subject matter content. In addition, resource persons should be limited to recognized experts to increase the validity and credibility of the program.

Involving a resource person in an adult education program requires a great deal of advance planning. Before contacting a potential resource person, the adult educator should have a clearly defined objective. The objective should be communicated to the potential resource person to ascertain his or her interest and ability to contribute to the fulfillment of the objective of the adult education program. Most resource persons welcome the guidance that a well-conceived objective provides as they prepare for their presentation or remarks. Adult educators should not be overly concerned about being too directive or restrictive by asking the resource person to focus on a pre-determined objective. This practice ensures that resource persons make a positive contribution to the overall objectives rather than pursuing another topic that may detract from the continuity of the program. Table 5 presents a number of guidelines to consider when using a resource person in an adult education program.

TABLE 5.

Using Resource Persons in Adult Education

Stage	Guidelines
Selecting the Topic and Resource Person	• Let the class or advisory committee identify problems to solve, questions to be answered, key points for the resource person to address. (Do not stop with just a general topic.) • Establish the objective for the program. • Have the advisory committee or planning group suggest individuals who would be possible resource persons.

(Continued)

TABLE 5. (Continued)

Stage	Guidelines
Preparing for the Program	• Contact resource person. Make sure details are clear, such as location, time, costs, etc. Confirm arrangements made by telephone with a letter. • Provide resource person with program objective, key points to be addressed, and specific questions to be answered. • Invite sponsor, if appropriate. • Ascertain A-V equipment or other materials and supplies needed for the resource person.
During the Program	• Review previous session. • State objective of the current session. • Set the stage for the resource person by briefly discussing problems, practices, participants' experiences related to the topic. • State the key points and/or questions the resource person will address. • Introduce resource person. Include facts, such as name, occupation, experiences, and training, relevant to the topic. • Allow participants to ask questions, and discuss how to use the information presented.
After the Program	• Summarize or conclude (briefly) the program with two or three guidelines for the adult participants to use in applying what was learned. • Send "thank you" letters to the resource person and the sponsor.

Summary

Effective adult educators should be recognized experts in their content area, in addition to having knowledge of teaching/learning principles, and the needs of the target audience. Part-time adult educators, who are usually employed in careers related to the content area, may need assistance in developing their teaching and facilitation skills. Adult educators also need to develop an awareness of the unique and diverse characteristics of adult learners. Professional development activities should be provided and encouraged to support adult educators in their content knowledge, teaching\learning practices, and understanding of adult learners.

DISTANCE LEARNING
WITH ADULTS*

Distance education is enjoying renewed popularity in adult education. This renewed interest has emerged, in part, because of advancements in communications and computer technology. Another reason for this resurgence may be due to changes in society and economics. Many adult learners simply cannot relocate to take advantage of educational opportunities offered by colleges and universities. Distance education addresses the issue of increasing access to educational opportunities for adults.

What Is Distance Learning?

The term *distance learning* is used to describe the intended outcome of the *distance education* process. Over the years, both terms have been used interchangeably with other words and phrases, such as outreach, extension education, tele-teaching, correspondence study, and independent study. For the purpose of clarity, this chapter will examine both the process and the outcome of distance education as they relate to adult audiences.

*W. Wade Miller, Professor, Agricultural Education and Studies, Iowa State University, Ames.

Distance education has been described in a number of ways. Dan Coldeway, of Athbasca University in Canada, provided a model to help describe the term by employing the variables of time and place (Simonson, 1995). Traditional, face-to-face, education happens at the same time and place. Distance education has often been described as taking place at different times and in different places. However, with advances in telecommunications technologies, distance education can happen at the same time, but in many different places.

Garrison and Shale (1987) proposed the following three criteria as essential in characterizing the distance education process:

1. Distance education implies that the majority of educational communication between (among) teacher and student(s) occurs noncontiguously;

2. Distance education must involve two-way communication between (among) teacher and student(s) for the purpose of facilitating and supporting the educational process; and,

3. Distance education uses technology to mediate the necessary two-way communication (p. 151).

Simonson and Schlosser (1995) incorporated much of Garrison and Shale's criteria into their definition of distance education. "Distance education implies formal institutionally-based educational activities where the teacher and learner are normally separated in location, but not normally separated in time, and where two-way interactive telecommunications systems are used for sharing video, data, and voice instruction (p. 153)."

Perhaps, the most concise definition was offered by Willis (1994). "At its most basic level, distance education takes place when a teacher and student(s) are separated by physical distance, and technology (i.e., audio, video, data, and print) is used to bridge the instructional gap (p. v)."

These recent definitions make provisions for the concept of time where communications between the student and the instructor can be at different times (asynchronous) or at the same time (synchronous). These definitions also reflect a belief that technology can be used to make the factor of location relatively unimportant to the teaching/learning process.

The Beginnings of Distance Education

Distance education for adults started more than 150 years ago. Its roots are in correspondence education for adult learners. Correspondence courses came about with the development of the postal system. One of the earliest references to correspondence education was an advertisement printed in a Swedish newspaper in 1833. The advertisement described a course in "Composition through the medium of the Post" (Holmberg, 1986, p. 6.). Printed materials and, later, photographic recordings were used in Europe during the 1800s.

Correspondence study found its way to the United States by 1873 (Watkins, 1991). Anna Eliot Ticknor founded a society to encourage study at home. The Society to Encourage Studies at Home involved more than 10,000 learners over a 24-year period. The learners were primarily women interested in classical studies. The learners corresponded monthly with their instructors who provided reading materials and exams.

Various forms of correspondence study became popular in the late 1800s. From 1883 through 1891, the Chautauqua College of Liberal Arts in New York offered academic degrees to learners who completed correspondence courses and on-site summer courses (Watkins, 1991). The university extension movement also embraced the correspondence concept for degree programs. Among the earliest in the extension movement were Illinois Wesleyan in 1877 and the University of Chicago in 1892.

Correspondence study has also been used in secondary schools, but the majority of efforts have been for adults. The target groups for correspondence study have generally been adults with occupational, social, and family commitments (Hanson, 1997). The target group today is still primarily adults. Distance education provides adults with the opportunity for intellectual pursuits as well as occupational knowledge and skills. Correspondence study provides individualized learning opportunities with flexibility as to time and place.

Correspondence study has adopted new forms of technology as it has become available. In the 1920s, there were at least 176 radio stations at educational institutions that transmitted correspondence study broadcasts (Buckland & Dye, 1991). By the 1950s, college-credit classes were being offered on a new media: broadcast television. Later on, video tape courses were developed. Audio recordings and laboratory kits have also been used for many years.

Educational television grew with the advent of satellite technology from the 1960s to the 1980s (Buckland & Dye, 1991). The first federally-funded experiment in satellite-delivered educational television was the Appalachian Educational Satellite Project in 1974–75. In 1988, Congress created the STAR Schools Program to develop multi-state, multi-institution, distance education programs for elementary and secondary schools (Willis, 1994). Partnerships between educational institutions and the communications industry were encouraged. Numerous projects benefiting both children and adults resulted from this program. Correspondence education continues to this day, but advances in technology are making profound changes in the way it is delivered to learners.

From One-Way to Two-Way Communication

In the late 1980s and early 1990s, advancements in fiber-optic communications technology have revolutionized distance education. The learner and the instructor are still separated by space, but not necessarily by time. Distance education can now be categorized in two ways: synchronous and asynchronous. Synchronous distance education takes place in real-time. The instructor and learner can engage in two-way communication during the learning session. Asynchronous distance education takes place at one time and is recorded for delivery at another time. Instructors and learners can communicate before or after the learning session orally or in writing via letter, FAX, email, etc.

Traditional correspondence education is asynchronous, but so are other forms of distance education, such as instructional television transmitted over the air, by cable, microwave, or satellite. Computer-aided instruction or computer-mediated instruction are also forms of asynchronous distance education. Computer programs are stored on electronic media, such as microcomputer disks, hard drives, and file servers. Sometimes, the computer programs are stored on a remote computer. The learner can establish a connection with the remote computer via telephone, computer modem, or through the Internet. In most cases, instructional programs provide for interaction between the learner and the program. In distance education, the learner interacts with the instructor in traditional ways (orally or in writing), but may use technology (telephone, electronic main, computer teleconferencing) to accomplish this task. Distribution of reading materials, assignments, and tests can be handled by postal or parcel service, facsimile, electronic mail, or via other electronic means.

Fiber-optic communications technology allows synchronous distance education programs to exist by linking distant locations. The technology provides for live, two-way, high-quality, audio and visual interactive systems. The learner and instructor are separated only by distance. Learners and instructors can see and talk to one another in much the same way as they would have in a traditional classroom. Distribution of materials are often handled in the same manner as asynchronous distance education.

Is Distance Education Effective?

Even with all the advancements in technology available to deliver distance education to adults, one question still remains: Is distance education as effective as traditional, face-to-face classroom instruction? This question was first asked 150 years ago and is still being asked today. Research comparing distance education to traditional education has shown that teaching and learning at a distance is as effective as traditional instruction, provided the methods and technologies utilized are appropriate to the instructional objectives and provisions are made to facilitate timely interaction among the learners and between the learners and the adult educator (Moore, et al., 1990). The effectiveness of distance education is strongly related to the interest of the adult learner in the subject matter being offered. *Active learning*, where the student interacts with the subject matter and finds ways to make the subject matter relevant, is highly encouraged in distance education. Adult learners have little tolerance for pre-recorded lectures, termed *talking heads*, with little or no provision for interaction.

Distance education is different from traditional face-to-face instruction in several ways. Communication between the learner and the instructor is very different. In a face-to-face situation, the instructor relies on many verbal and non-verbal cues from the learner. For example, the instructor can see if the students are taking notes, if they appear interested, observe their facial expressions, and determine if someone wants to make a comment or ask a question. The instructor can make adjustments to the process, as well as to the content of the lesson, to meet the needs of the class.

In distance education situations, the instructor may not be able to receive visual cues. If these cues exist, they are transmitted through cameras, transmission lines, and finally a television monitor. Much of the visual information is lost, especially if a student is sitting in the back of a room or the lighting is inadequate. Without visual cues, the instructor does not know if the learners are attentive, if they are talking among themselves, or even if they have left the room. It is more difficult to answer questions or to offer additional explanations to the learners. Students at different down-link sites also find it more challenging to develop a rapport with their colleagues.

To ensure that distance education programs are effective, the instructor should consider the following factors: understanding the distance learner, planning and organizing the program and content, planning for interaction and feedback, using effective teaching strategies, and evaluating learner performance.

Understanding the Distance Learner

Distance learners may be different than face-to-face learners. It is important for the adult educator to remember that distance learners as a group are more heterogeneous than traditional face-to-face learners (Holmberg, 1986). Distance education serves learners who cannot or will not make use of face-to-face learning opportunities. Distance learners are often "place-bound." Attending face-to-face classes or meetings may be difficult or impossible for them due to employment, family, community, or financial considerations. Some learners do not want to sit in a classroom and learn the traditional way, even if convenience is not a barrier.

Distance learners are often more independent. They will usually take a more active responsibility for their own learning. Self-discipline is important because distance learners have to make learning a high priority in their lives. Some adult learners fail to make this commitment resulting in higher non-completion rates when compared to on-site adult learners. Distance education promotes learners' freedom of choice and independence. The adult educator must recognize these phenomena when planning distance education offerings for adults.

The adult educator should not assume that the distance learners all have the same educational goals. Many distance education programs are designed around a degree program, but some distance learners will be participating in the program for personal development or for professional development objectives. These students may have different expectations regarding the outcome of distance education offerings (Willis, 1993).

Planning and Organizing the Program Content

The quality and quantity of content in a distance education offering should not be reduced. Some adult educators argue that it takes more time to teach via distance education and, therefore, they cannot "cover" as much material. While this may be true if the instructor presents all the content to the learner orally, it is not true if the instructor provides the content to the learner by alternate means. Content can be planned, delivered, and stored in several ways. Course materials, including lectures and reading materials, can be distributed in print form or stored on computer media, such as CD-ROM's. Course materials can also be stored on a file server and accessed by learners via computer modem. Study guides, handout materials, and notes pages should be developed and distributed before a learning session begins. Recorded lectures should be brief and contain high quality visuals to be effective. The instructor should consider distributing all course materials before the class begins, because adult learners will make use of the materials at different rates of time. Distance learners also want to know what will be expected of them in advance of the course.

The adult learner may not have access to the reading materials required in a course. They may have access only at a local public library. It will be difficult for them to search for literature to prepare reports and papers. The adult educator will need to provide access to the appropriate literature. However, it is necessary to follow fair-use guidelines in duplicating and distributing copyrighted materials.

Plan for Interaction and Feedback

Acker and McCain (1993) stated that "interaction is central to the social expectations of education in the broadest sense and is in itself a primary goal of the larger educational process and that feedback between the learner and teacher is necessary for education to develop and improve" (Acker & McCain, 1993, p. 11.). Moore (1989) further defined interaction by dividing it into three categories: learner-content interaction, learner-instructor interaction, and learner-learner interaction. A fourth component of interaction was identified by distance education researchers as learner-interface interaction that takes into account the interaction that occurs when a learner must use intervening technologies to communicate with the content, negotiate meaning, and validate knowledge with the instructor and other learners (Hillman et al., 1994).

In asynchronous distance education situations, it may appear that interaction is not possible. However, interaction or the perception of interaction, can and should be a planned part of instruction (Miller, Doerfert, & King, 1996). The perception of interaction can be created by "personalizing" the course through the use of learner names, photos, and materials submitted by the learners. If instruction is prepared in advance and delivered in a one-way audio-video format at another time, such as broadcast television or videotape courses, interaction can be incorporated in several ways. One way is to require learners to phone questions or comments in to an answering machine. The learner can record an audio or video tape and send it to the instructor. The instructor can replay selected recordings in subsequent class sessions and respond to these questions. The instructor can show a photo on the screen of the learner who asked the question. The instructor could have one or more learners present and interact as a part of the program. One or more live group meetings with adult learners and the use of the telephone are advisable. The key is to involve learners as much as possible during the program.

Interaction in a course or program on the Internet is also possible. Learners will interact with the computer, but they can also interact with other learners as well as the instructor. The use of electronic mail or web conferencing software makes interaction relatively easy. The instructor should be an active participant in the discussion and should encourage participants to post comments and questions.

Interaction in synchronous distance learning sessions is relatively easy to accommodate. After all, two-way, full motion, compressed video systems are designed to be interactive. The instructor can see and hear the students (although maybe not all the students at the same time), and the students can see and hear the instructor. Questions, comments, and discussion can take place much as it does in a traditional classroom. However, instructors soon discover that this is not the case. There typically is less interaction when in this type of distance education setting. In most cases, the learner must press a button on a microphone to ask a question. When the person does this, the audio control from the instructor is interrupted and transferred to the learner. Some people are reluctant to push the button. In addition, there may be a feeling of isolation between the instructor and learner because the interaction must take place through a conduit of communication technology. To overcome this situation, the instructor should take the necessary steps during the first session to help the learners become thoroughly familiar with the use of the equipment. It is important to have learners practice asking questions over the system. When teaching, the instructor may have to call on individual participants to seek their input. It is advisable to require students to provide a photograph so the instructor can connect students' names with their faces. The instructor should be aware that it takes about three times the amount of time to ask questions and receive answers from distance learners. Instructors can build in group discussions or projects with distance learners and require reports to be given over the system.

Interaction and feedback should be planned. The instructor should consider the following strategies to improve interaction (Willis, 1993):

- Use a variety of delivery systems for interaction, including individual telephone calls, conference calls, facsimile, electronic mail, and computer conferencing.

- Contact each site weekly, especially, early during the course. Contact missing learners.

- Write comments on assignments and other materials submitted by the learners. Return assignments promptly using facsimile or electronic mail, if possible.

- Establish office hours and provide learners with a telephone number. Consider obtaining a toll-free number. Use electronic mail when available.

- Early in the course, require learners to contact him/her via electronic mail or telephone.

- Prepare study questions or notes pages for use by learners.

- Consider having learners keep a journal and submit entries periodically.

- Contact individual learners outside class time to solicit feedback and check progress.

- Use an on-site facilitator to stimulate interaction and serve as a link between the instructor and the learners.

- Use effective teaching skills.

Using Effective Teaching Strategies

Effective distance teaching does not require the instructor to learn a new set of skills, rather, it requires the instructor to modify existing behaviors. First, instructors should rethink their programs. In many cases, it will be necessary to redesign the program in such a way that the content is largely transferred to the learner in a form other than lecture. The instructor may need to identify and supply textbooks, reference materials, supplies, other reading materials, audio or visual materials, or other resources for learner use. Teaching outlines, plans, visual aids, demonstrations, and class activities will all have to be examined for suitability in the distance education setting. For example, it may be necessary to send a box of materials to a distance education site if the instructor wants the learners to conduct an experiment or perform a class activity.

The role of the instructor will need to switch from that of an "expert" content provider to that of a learning facilitator. The learners must assume more responsibility for learning the content, and the instructor will help them understand the content and its relevance.

The instructor must become proficient in the use of the technology that is being used. Training on the use of the system is imperative. The instructor will also have to become thoroughly familiar with the strengths and limitations of electronic delivery systems. Visual aids, such as slides and transparencies, may have be reformatted. As much as 75 percent of the image resolution of a computer generated visual is lost when transmitted through a television monitor. Schmidt (1997) provides some general guidelines for preparing visuals to be used in electronic presentations:

- Keep the composition of the visual brief and to the point. Use key words or phrases, not sentences.

- Do not crowd words on a page. A good rule of thumb is five words per line and five lines per page. Leave plenty of margin space all around the visual.

- Avoid colors that do not display well on television. These include red and orange. The goal is to provide adequate contrast between text color and background color.

- Consider using light-colored letters against a darker-colored background, such as blue or green. Do not use a white background. Limit colors to two or three.

- Use a plain type style, without serifs, of adequate size. It may be necessary to use bold face type.

- Start off the main body of the visual with 36 point and adjust up or down as needed. The minimum size is considered to be 18 point.

- Consider using subtle transitions and builds on slides and make generous use of clip art, pictures, graphs, and diagrams.

- When using 35mm slides or transparencies, always use horizontal compositions. Remember, television is formatted in a 3x4 ratio.

Evaluating Learner Performance

One of the first questions adult educators ask when planning to teach a distance education course is, "How will I administer quizzes and exams?" The task of evaluating the performance of distance learners is a more complex procedure when administering exams. It also includes how to accomplish the other assignments educators typically have in a class, such as oral and written reports, research papers, case studies, reviews of literature, and group projects. The adult educator will need to address the issue of academic honesty on the part of adult learners. If the adult educator accepts the position that adult learners have the responsibility to learn and apply the material offered in a distance education offering, then the task of evaluating them is much easier. On the whole, adult students have a great amount of academic integrity. They will generally follow the guidelines and policies established by the instructor. "Grading" is not a large issue; their learning and understanding of the material are what is important. With this in mind, many adult educators reduce or eliminate the use of quizzes and exams in their programs and increase the number or complexity of writing assignments where learners can demonstrate their knowledge and understanding of the concepts being presented.

Administering quizzes and exams in an asynchronous distance education situation can take several forms. Exams can be sent to a person who will proctor the exam session. The proctor can distribute the exams to the learners and return them to the instructor. Exams can also be administered orally over the telephone. Another way is to distribute exams by electronic mail and have the learners submit the completed exams to the instructor the same way. There are also computer software programs designed for this purpose.

Administering exams in a synchronous distance learning situation can take place much as it does in a face-to-face classroom. Quizzes and exams can be sent to the site in advance. A proctor, site facilitator, or trusted participant can administer the exams and return them to the instructor. Quizzes and exams can be shown on visuals to all sites. Learners can respond to the questions and send their answers to the instructor by mail or facsimile.

Making assignments of papers and reports is another issue in learner evaluation. Submitting a written assignment to the instructor is easy; the difficultly comes in making sure the learner has all the necessary resources to complete the assignment. The instructor will need to take more responsibility for supplying the necessary resources as the adult learner may not have access to the materials. The instructor must take responsibility for securing permission to use copyrighted materials and following fair use guidelines. Oral reports are possible in asynchronous learning situations by using audio cassette recordings or even short answering machine messages. Oral reports in synchronous distance learning situations can take place much as they do in face-to-face classrooms.

No matter what evaluation procedures are employed by the instructor, it is very important to supply the adult learners with timely feedback. Adult learners feel left out or forgotten when they do not hear back from the instructor for long periods of time. The instructor should consider using electronic mail, the facsimile, and the telephone as well as the postal system to provide feedback to the participants.

Summary

Distance education was started to serve adult learners' needs more than 150 years ago with the introduction of the postal system. Many adults are "place-bound" due to a number of factors relating to community and family obligations, socio-economic status, and employment considerations. Distance education is growing and changing to meet the needs and demands of adults in a variety of situations. Adults need access to high-quality, relevant distance education. Advancement in communications and computer technology is making distance education more available, more convenient, less costly, and more effective. The challenge for adult educators is to learn how best to take advantage of new technologies to produce high-quality educational offerings that serve the needs of adult learners.

A PHILOSOPHY FOR ADULT EDUCATION*

It is important to preserve the uniqueness of individuals and groups within a reasonably disciplined social context. Doing so allows for and promotes ways of preserving the differences we need to live and grow.

Adult education has the important general purpose to discover and present to the adult the opportunity to advance as a maturing individual.

This philosophy points toward the use of adult education for the development of free, creative, and responsible persons in order to advance the human maturation process.

The philosophy of adult education is based on the belief that:

1. Adult behavior can be changed to some extent.

2. Adult education should be designed to help people grow and mature.

3. Adults must be offered and helped to use the opportunity to act responsibly in several facets of their lives: political, vocational, cultural, spiritual, and physical.

---

*Adapted from Bergevin, P. (1967). *A Philosophy for Adult Education*. New York: The Seabury Press, pp. 3-5.

4. Adults should assume the obligation to learn to become more productive citizens.

5. Adults have untapped resources of creative potential that should be utilized.

6. Every conscious adult can learn.

7. All adults can be helped to make better use of their intellectual capacity.

8. Adults need to live together in a community to grow and mature, and they need to learn how to do this.

9. All adults should find some way to express themselves constructively and creatively.

10. Traditional teaching procedures and learning facilities are often inadequate.

11. An understanding of freedom, discipline, and responsibility promotes the discovery and productive use of our talents.

12. Such vital concepts as freedom, discipline, and responsibility can be comprehended by experiencing them through a variety of inspired learning experiences in a host of subjects.

13. What is called a free or democratic society must strongly emphasize lifelong learning for all its citizens if they propose to remain free and use their freedoms effectively.

14. Each adult participating in a learning experience should have the opportunity to help diagnose, plan, conduct, and evaluate that experience along with his/her fellow learners and administrators.

15. The civilizing process is evolutionary and will advance in proportion to the number and intellectual quality of the adults who play an active role in that process.

16. Many adults associate education with a school. Education can take place at home, in church, in a factory, on a farm, in many different places.

17. The means are as important as the ends.

18. A human being is neither "good" nor "bad," but essentially an adaptable, educable person in a state of becoming, as well as being, and capable of a degree of excellence he/she rarely attains.

19. Behavior is conditioned by feelings and emotions as well as by reason and rational judgment.

20. Human beings seek fulfillment or happiness.

21. Adult education can help condition persons to live in a society and at the same time sensitize them to ways in which that society can be improved.

22. Up to the present, the democratic idea has seemed to fit the nature of human beings and also adult education.

APPENDIX B

30 THINGS WE KNOW FOR SURE ABOUT ADULT LEARNING*

1. Adults seek learning experiences in order to cope with specific life-changing events.

2. The more life-changing events adults encounter, the more likely they are to seek learning opportunities.

3. The learning experiences adults seek on their own are directly related — at least in their own perceptions — to the life-change event(s) that triggered the seeking behavior.

4. Adults are generally willing to engage in learning experiences before, after, or even during the actual life-change event.

5. For most adults, learning is not its own reward; they have use for the knowledge or skill being sought. Learning is a means to an end, not an end itself.

6. Increasing or maintaining one's sense of self-esteem and pleasure are strong secondary motivators for engaging in learning experiences.

7. Adult learners tend to be less interested in survey courses.

*Adapted from Zemke, R. & Zemke, S. (1981). *Training, 18*, p. 45-49.

8. Adults need to be able to integrate new ideas with what they already know.

9. Information that conflicts with what adults believe to be true is integrated more slowly.

10. Information that has little "conceptual overlap" is acquired slowly.

11. Fast-paced, complex, or unusual learning tasks interfere with learning.

12. Adults compensate for being slower in psychomotor learning tasks by being more accurate and by making fewer trial-and-error mistakes.

13. Adults tend to take errors personally and are more likely to let the error affect their self-esteem.

14. Curriculum designers must know whether the concepts or ideas will be in concert or in conflict with learner and organizational values.

15. Programs need to be designed to accept viewpoints from people in different life stages and with different value "sets."

16. A concept needs to be anchored or explained from more than one value set and appeal to more than one developmental stage.

17. Adults prefer self-directed and self-designed learning projects over group learning experiences led by a professional.

18. Non-human media, such as books, programmed instruction, and television, have become popular in recent years.

19. Regardless of media, straightforward how-to is the preferred content orientation.

20. Self-direction does not mean *in isolation*.

21. The learning environment must be physically and psychologically comfortable.

22. Adults have something real (self-esteem and ego) to lose in a classroom situation.

23. Adults have expectations; and it is critical to take time up front to clarify and articulate all expectations before getting into content.

24. Adults bring a great deal of life experience into the classroom, an invaluable asset to be acknowledged, tapped, and used.

25. Instructors who have a tendency to hold forth rather than facilitate can hold that tendency in check (or compensate for it) by concentrating on open-ended questions to draw out relevant trainee knowledge and experience.

26. New knowledge has to be integrated with previous knowledge through active participation.

27. The key to the instructor role is to control the pace of the course through the introduction of new material, discussion, sharing, and the clock.

28. The instructor has to protect minority opinion.

29. Integration of new knowledge and skill requires transition time and focused effort.

30. Learning and teaching theories function better as a resource than as a Rosetta stone.

PRINCIPLES OF TEACHING AND LEARNING*

Principle 1   When the subject matter to be learned possesses meaning, organization, and structure that is clear to the learners, learning proceeds more rapidly and is retained longer.

Principle 2   Readiness is a prerequisite for learning. Subject matter and learning experiences must be provided that begin where the learner is—that is, at the learner's current level of knowledge.

Principle 3   Learners must be motivated to learn. Learning activities should be provided that take into account the wants, needs, interests, and aspirations of the learners.

Principle 4   Learners are motivated through their involvement in setting goals and planning learning activities.

Principle 5   Success is a strong motivating force.

---

*Adapted from Newcomb, L. H., McCracken, J. D., Warmbrod, J. R. (1993). *Methods of Teaching Agriculture* (Chapter 2). Danville, IL: Interstate Publishers, Inc.

Principle 6   Learners are motivated when they attempt tasks that fall in a range of challenge such that success is perceived to be possible, but not certain.

Principle 7   When learners have knowledge of their learning progress, performance will be superior to what it would have been without such knowledge.

Principle 8   Behaviors that are reinforced (i.e., rewarded) are more likely to be learned.

Principle 9   To be most effective, reward (reinforcement) should follow as immediately as possible the desired behavior and be clearly connected with the behavior by the learner.

Principle 10   Directed learning is more effective than undirected learning.

Principle 11   To maximize learning, learners should "inquire into" rather than be "instructed in" the subject matter. Problem-oriented approaches to teaching improve learning.

Principle 12   Learners will learn what they practice.

Principle 13   Supervised practice is more effective when it occurs in a functional educational environment.

VARIABLES OF
EFFECTIVE TEACHING*

Clarity

Use clear explanations, examples, and assignments; use terms understood by the learners; answer questions directly; follow an organized approach to the subject.

Variability

Use a variety of instructional materials and learning activities; vary the method of evaluation and assessment; vary the level of interaction between you and the learners.

Enthusiasm

Demonstrate a "passion" for the subject matter through the use of movement, gestures, facial expressions, voice inflection, etc.

Task-Oriented and Businesslike Behavior

Demonstrate concern that the learners gain something of value through each learning activity; encourage learners to work hard through independent and creative effort.

---

*Rosenshine, B. & Furst, M. (1971). Research on teacher performance criteria. In B. O. Smith (Ed.)., *Research in teacher education,* (pp. 37-72). Englewood Cliffs, NJ: Prentice-Hall.

Opportunity to Learn Criterion Material	Provide opportunities for learners to learn important concepts through activities and assignments; be sure important concepts are reflected in assignments and evaluation activities.
Use of Learner Ideas and General Indirectedness	Utilize input from learners in group discussion; praise and encourage ideas provided by learners to gain the learners' perspective of the concepts.
Criticism	Maintain high correlation between teacher criticism and learner performance; inform learners when they are wrong without personal criticism.
Use of Structuring Comments	Use statements designed to provide cognitive scaffolding by linking new experiences with past experiences; review key points before each session and summarize after.
Types of Questions	Ask questions at various cognitive levels: "What?," "Where?," and "When?" require lower-level thinking; "Why?" and "How?" stimulate higher-level thinking.
Probing	Respond to student answers by asking for elaboration (e.g., "What factors led you to that conclusion?").
Level of Difficulty of Instruction	Project an image that the subject matter is neither too hard nor too easy for the learners to grasp—that success is possible but not guaranteed.

APPENDIX E

COMPARISON BETWEEN PEDAGOGY AND ANDRAGOGY

Regarding:	Pedagogy (helping children learn)	Andragogy (helping adults learn)
Concept of the Learner	• Dependent • Educator is responsible for content and evaluation	• Self-directed • Educator is to guide, nurture, and encourage
Role of the Learner's Experience	• Limited experience • Most experience is from prior educational settings	• Rich resource for learning • Encourage sharing of experiences with others
Readiness to Learn	• Maturity dictates readiness • Age is a major determinant of readiness	• Motivated to learn only when the need is evident • Must see a practical application
Orientation to Learning	• Education involves acquisition of content knowledge and facts • Little relation is evident between subjects	• View education as a process of developing to fullest potential • Focus on application and solutions to problems

LEVELS OF LEARNING WITHIN DOMAINS

The taxonomy of educational objectives classifies objectives by complexity as well as by domain. The taxonomy assumes that cognitive learning is hierarchical in nature and proceeds in an orderly progression. Both the affective and the psychomotor domains form a continuum. The degree to which values are internalized is the basis for the affective continuum. The psychomotor continuum involves the degree of motor-skill development.

COGNITIVE DOMAIN

Bloom, et al. (1956) identified six levels of learning in the cognitive domain.

Knowledge	The learner remembers facts, generalizations, methods, processes, and criteria in a form similar to that in which they were studied.
	Characteristic behaviors include: recall, recite, list, define, and tell.
Comprehension	The learner can make some use of the material in a form other than originally studied.
	Characteristic behaviors include: explain, interpret, give examples, and translate.

Application   The learner can use knowledge (facts, generalizations, methods, processes, and criteria) to solve an unfamiliar problem or in a new situation.

Characteristic behaviors include: apply, solve, use, employ, compute, show, and construct.

Analysis   Learners can break down an idea, method, or design into its component parts.

Characteristic behaviors include: discriminate, contrast, compare, relate, categorize, distinguish, and recognize assumptions. Comprehension level objectives are sometimes mistaken for analysis.

Synthesis   The learner draws together knowledge and ideas from multiple sources to produce a unique product, design, or plan.

Characteristic behaviors include: design, create, propose, plan, arrange, modify, adapt, and compose.

Evaluation   The learner can make judgments of worth using some set of internal or external criteria.

Characteristic behaviors include: judge, rate, appraise, evaluate, choose, and assess.

AFFECTIVE DOMAIN

The affective domain describes feelings, values, and attitudes.

Receiving (Attending)

Responding   The learner does something about selected stimuli; he or she complies with request or directions.

Characteristic behaviors include: is willing to, complies with.

Valuing	The learner sees worth or value in a thing.
	Characteristic behaviors include: actively participates.
Organization	The learner begins building a value system, which places values in a hierarchy.
	Characteristic behaviors include: forms judgements, seeks to analyze, weighs alternatives, and clarifies values.
Characterization	At this stage, the learner has a well-established value system that guides behavior.
	Characteristic behaviors relate to consistency of action in regard to the value system, philosophy of life, or personal code.

The affective domain is more abstract than the cognitive domain. The personal nature of values and value systems makes setting instructional goals more difficult.

PSYCHOMOTOR DOMAIN

When the original committee decided against developing the third handbook on the taxonomy, Elizabeth Simpson (1967) assumed the task of identifying and describing the classification system for the psychomotor domain. As a home economist and vocational educator, Simpson recognized the need for development of this domain for home economics as well as for other fields.

Perception	The learner becomes aware of objects, qualities, or relationships by way of the sense organs and selects and translates cues.
	Characteristic behaviors include: hears, sees, feels, tastes, and smells in order to select and determine meaning of cues. These behaviors are directed toward learning a psychomotor task.

Set

The learner adjusts or remains in readiness for a particular kind of action or experience. Three aspects of set are mental, physical, and emotional.

Characteristic behaviors include: is mentally ready, assumes bodily stance, positions hands, and is ready to respond.

Guided Response

The learner performs the overt behavior under the instructor's guidance. This category includes imitation as well as trial and error.

Characteristic behaviors include: imitates, tries various responses, and discovers.

Mechanism

The student has learned a habitual response, which he/she performs with a degree of confidence and skill. The act is part of a repertoire of possible responses and may involve a pattern of responses in carrying out the task.

Characteristic behaviors include: perform and mix. This level is distinguished from guided response in that the learner needs no prompting in selecting and performing the act.

Complex Overt Response

The learner can perform a motor act that requires a complex movement pattern smoothly and skillfully. This stage includes resolution of uncertainty and automatic performance.

Characteristic behaviors include: demonstrates skill, proceeds without hesitation, and controls actions with ease (Simpson, 1967).

APPENDIX

CONTINUING EDUCATION UNITS*

One of the major branches in the field of adult education is Continuing Professional Education (Rachal, 1987). Programs of Continuing Professional Education are planned and conducted to maintain and enhance the competence of persons engaged in a wide range of occupations or professions. Medicine, law, nursing, teaching, and accounting are examples of professional fields that require practitioners to complete continuing education activities to maintain professional licensure or certification in their respective field.

The Continuing Education Unit (CEU) was developed in 1970 in an attempt to provide a "standard" form of credit for continuing education or training. A national task force developed the following definition in 1970 as an initial step to standardize the awarding of credit for con-

---

*For additional information about awarding CEU credit for adult education programs contact:

International Association for Continuing
Education and Training
1200 19th Street, NW
Suite 300
Washington, DC 20036-2412

(202) 857-1122 Phone
(202) 223-4579 FAX

tinuing professional education. The CEU was defined as (Phillips, 1994, p. xi):

Ten contact hours of participation in an organized continuing education experience under responsible sponsorship, capable direction, and qualified instruction.

Since 1970, thousands of sponsors in the United States have used this definition as the standard for awarding Continuing Education Units. Some continuing professional education programs award credits based on contact hour (50 to 60 minutes). Program sponsors are free to decide which unit(s) of measure they will use to award credit, whether to test participants, or to use attendance as the criteria for successful completion and the awarding of CEU credit.

CRITERIA FOR CONTINUING EDUCATION UNIT

In 1972, the International Association for Continuing Education and Training established the following criteria as program standards to qualify for awarding Continuing Education Unit (CEU) credit (Phillips, 1994, p. xiv):

1. Each activity is planned in response to educational needs that have been identified for a target audience.

2. Each activity has a clear and concise written statement of intended learning outcome.

3. Qualified instructional personnel are involved in planning and conducting each activity.

4. Content and instructional methods are appropriate for the intended learning outcome.

5. Participants must demonstrate their attainment of the learning outcome.

6. Each learning activity is evaluated by the participants.

7. The sponsor has identified clearly defined unit, group, or individual responsibilities for developing and administrating learning activities.

8. The sponsor has a review process in operation that ensures the CEU criteria has been met.

9. The sponsor maintains a complete record of each individual's participation and can provide a copy of that record upon request for a period of at least seven (7) years.

10. The sponsor provides an appropriate learning environment and support services.

PROMOTION OF ADULT LEARNING

Common Promotional Materials

Newspaper Advertising	• Aim to the target audience • Select sections of the newspaper that reach the target audience • Numerous small ads are more effective than fewer large ads • Ads need to be eye-catching, easy-to-read, and dignified • Relate ads to specific activities • Seek professional advice from newspaper advertising and marketing representatives
Newspaper Publicity	• Note: Publicity is free; advertising is paid for • Develop good relationships with the media • Invite the press to newsworthy activities • Custom write news releases to the paper's specifications • Do not favor one newspaper over another
Radio and Television	• Aim to attract an audience not reached by other media • Principles that apply to news articles also apply to radio and television

(Continued)

Common Promotional Materials (Continued)

Direct Mail and Printed Materials (including folders, bulletins, booklets, and flyers)	• Use to follow-up on awareness created by mass media • Should be more personalized than mass media messages • Most program participants are obtained through direct mail • Aim is to tell important information about the program • Quality should be "first rate" with an attractive cover • Information should include program title, description, dates, time, place, and special qualifications of instructors • Flyers can be used to promote one or a group of programs and tailored to attract specific audiences
Form Letters	• Always use program or institutional letterhead • Can be used as a cover letter for bulletins, flyers, booklets, etc. • Can be used to highlight items of interest to specialized groups • Can be used as separate mailing pieces targeted to specific groups
Newsletters	• Used to convey activities, schedule of visits, and program schedules • Distribution on a regular and timely basis is important • Keep mailing list current and accurate
Posters, Displays, Exhibits, Open House, and Tours	• Excellent ways to attract attention and new program prospects • Keep sign-up sheets of attendees to add to a mailing list • Video tape presentations of previous program activities can be used to stimulate interest

Promotional Media Advantages and Limitations*

Strategy	Advantages	Limitations
Press Release (news)	• Reaches wide circulation • Free publicity • Press coverage lends credibility	• Not good for limited/small audience • May not be best method for reaching target audience • Time of day (newscast) • Page on which article appears (newspaper) • Length of story/article affects audience
Public Service Announcement (PSA) (radio-TV)	• "Free ads" on the air • Good tool for public education programs	• May be aired at odd hours/times • Ad often produced cheapest way possible • Pre-recorded PSAs must meet quality standards of station
Free Speech Message (FSM)	• Free • Able to say anything about an issue without censorship	• Usually is produced inexpensively and may be of poor quality • May not be aired at time to reach intended audience
Calendar Listing	• Good for reminding audience of date, time, and place of events	• Primarily used for event publicity, not for PR • Provides only who, what, when, and where • May not be seen
Interview Show	• Free publicity • Allows speaker to clarify issues in-depth • Provides a public forum for issues • Allows person to speak for himself or herself, rather than relying on a reporter's interpretation	• Limited audience interest • Usually produced inexpensively • May appear to be a "talking head" • People may "tune out" after brief time
Slide-Tape, Video	• Visual presentation of issues, facts, resources • Good stimulus for discussion • Experiential (visually) • Adds variety and interest	• Possibly expensive • Need people to present tape • Need equipment • Need expertise to produce material

(Continued)

Promotional Media Advantages and Limitations (Continued)

Strategy	Advantages	Limitations
Banner	• Eye catching • Adds to feeling of event or festivity • Creates sense of community involvement	• Must be delivered and removed at certain times • Community regulations may prohibit • Expensive if commercially produced
Bus Poster	• Good for reaching a commuter audience • Targets youth and elderly	• May be expensive (cost per bus plus printing) • Limited audience exposure
Public Service Announcement (PSA) (print media)	• May attract more attention than printed article • Good supplement to article for event publicity	• Must run a week or more to be effective
Event or Action	• Good chance of getting coverage • Attracts public attention • Could be educational or entertaining • Brings issues to public attention • Allows direct contact with public • Creates opportunity for media follow-up	• Requires planning • Time consuming • Special materials may be required • Requires pre-publicity
Press Packet	• Gives media background information for stories • Could lead to more in-depth coverage	• Needs to be updated
Newsletter	• Good for creating network • Allows for in-depth information • Provides forum for target audience • Provides a clearinghouse for information	• Costly • Time consuming • Requires submissions from participants • Deadlines • Difficult to keep interesting and current • Special equipment may be needed

*Adapted from material prepared by Joan-Patricia O'Connor, O'Connor PR & Marketing.

REFERENCES

Acker, S. R. & McCain, T. A. (1993). *The contribution of interactivity and two-way video to successful distance learning applications: A literature review and strategic positioning.* Columbus: The Center for Advanced Study in Telecommunications. The Ohio State University.

Apps, J. W. (1991). *Mastering the teaching of adults.* Malabar, FL: Krieger Publishing Co.

Bandura, A. (1986). *Social foundations of thought and action.* Englewood Cliffs, NJ: Prentice-Hall.

Bankirer, M. W. (1995). Adjunct faculty as integrated resources in continuing education. In V. W. Mott & L. C. Rampp (Eds.). *Organization and Administration of Continuing Education* (pp. 135-143). Checotah, OK: AP Publications.

Bergevin, P. (1967). *A philosophy for adult education.* New York: The Seabury Press.

Bode, H. B. (1929). *Conflicting psychologies of learning.* New York: Heath.

Boone, E. J. (1985). *Developing programs in adult education.* Englewood Cliffs, NJ: Prentice-Hall.

Broomall, J. K. & Fisher, R. B. (1995). Budgeting techniques in continuing education. In V. W. Mott & L. C. Rampp (Eds.). *Organization and Administration of Continuing Education* (pp. 277-301). Checotah, OK: AP Publications.

Buckland, M. & Dye, C. M. (1991). *The development of electronic distance education delivery systems in the United States. Recurring and emerging themes in history and philosophy of education.* (ERIC Document Reproduction Service No. 345 713).

183

Caffarella, R. S. (1994). *Planning programs for adult learners*. San Francisco: Jossey-Bass.

Cherem, B. (1990, September). Adult education: A searching stepchild. *Adult Learning*, 23-26.

Covey, S. R. (1989). *The 7 habits of highly effective people*. New York: Simon & Schuster.

Cross, K. P. & Valley, J. R. and Associates. (1974). *Planning non-traditional programs: An analysis of the issues for postsecondary education*. San Francisco: Jossey-Bass.

Daines, J., Daines, C., & Graham, B. (1993). *Adult learning: Adult teaching*. England: University of Nottingham, Department of Adult Education.

Darkenwald, G. G. & Merriam, S. B. (1982). *Adult education: Foundations of practice*. New York: Harper & Row.

Dean, G. J. (1994). *Designing instruction for adult learners*. Malabar, FL: Krieger Publishing Co.

Delahaye, B. L., Limerick, D. C., & Hearn, G. (1994, Summer). The relationship between andragogical and pedagogical orientations and the implications for adult learning. *Adult Education Quarterly, 44*(4), 187-200.

Dewey, J. (1938). *Experience and education*. New York: Collier.

Dobbs, R. C. (1970). *Adult education in America: An anthological approach*. Cassville, MO: Litho Printers.

Galbraith, M. W. (1989). Essential skills for the facilitator of adult learning. *Lifelong Learning: An Omnibus of Practice and Research, 12*(6), 10-13.

Galbraith, M. W. (Ed.). (1991). *Facilitating adult learning: A transactional process*. Malabar, FL: Krieger Publishing Co.

Garrison, D. R. & Shale, D. (1987). Mapping the boundaries of distance education: Problems in defining the field. *The American Journal of Distance Education, 1*(1), 7-13.

Grow, G. O. (1991). Teaching learners to be self-directed. *Adult Education Quarterly, 41*(3), 125-149.

Hanson, D. (1997). Distance education: Definition, history, status, and theory. *Encyclopedia of Distance Education Research in Iowa* (2nd Edition). Ames, IA: Teacher Education Alliance.

Hillman, D. C., Willis, D. J., & Gunawardena, C. N. (1994). Learner-interface interaction in distance education: An extension of contemporary models and strategies for practitioners. *The American Journal of Distance Education, 8*(2), 31-42.

Holmberg, B. (1986). *Growth and structure of distance education*. London: Croom Helm.

Imel, S. (1994). *Guidelines for working with adult learners*. (Digest no. 154.) ERIC Clearinghouse on Adult, Career, and Vocational Education. Columbus: The Ohio State University, Center on Education and Training for Employment.

Jones, H. E. & Conrad, H. S. (1933). The growth and decline of intelligence. *Genetic Psychology Monographs, 13*, 223-298.

Kahler, A. A., Morgan, B., Holmes, G. E., & Bundy, C. E. (1985). *Methods in adult education* (4th Edition). Danville, IL: The Interstate Printers & Publishers, Inc.

Kerka, S. (1994). *Self-directed learning: Myths and realities*. ERIC Clearinghouse on Adult, Career, and Vocational Education. Columbus: The Ohio State University, Center on Education and Training for Employment.

Kirsch, I. S., Jungeblut, A., Jenkins, L., & Kolstad, A. (1993). *Adult literacy in America*. Washington, DC: U.S. Department of Education, Office of Educational Research and Improvement, National Center for Education Statistics.

Knoll, J. H. (1983). *Motivation for adult education*. Working papers presented to the European Conference on Motivation for Adult Education. Hamburg, Germany.

Knowles, M. S. (1962). *The adult education movement in the United States*. New York: Holt, Rinehart, and Winston, Inc.

Knowles, M. S. (1980). *The modern practice of adult education: From pedagogy to andragogy*. New York: Cambridge.

Knox, A. (Ed.). (1980). *Teaching adults effectively*. New Directions for Continuing Education, No. 6. San Francisco: Jossey-Bass.

Knox, A. B. (1980). *Developing, administering, and evaluating adult education*. San Francisco: Jossey-Bass.

Kolb, D. A. (1984). *Experiential learning*. Englewood Cliffs, NJ: Prentice-Hall.

Kopka, T. L. C. & Peng, S. S. (1993). *Adult education: Main reasons for participating*. Washington, DC: U.S. Department of Education, Office of Educational Research and Improvement, National Center for Education Statistics.

Kopka, T. L. C. & Peng, S. S. (1994). *Adult education: Employment-related training*. Washington, DC: U.S. Department of Education, Office of Educational Research and Improvement, National Center for Education Statistics.

Korb, R., Chandler, K., & West, J. (1991, September). *Adult education profile for 1990-91*. (NCES 91-222). Washington, DC: U.S. Department of Education, Office of Educational Research and Improvement, National Center for Education Statistics.

Kowalski, T. J. (1988). *The organization and planning of adult education*. Albany: State University of New York Press.

Lindeman, E. (1926). *The meaning of adult education*. New York: New Republic.

Long, H. B. (1983). *Adult learning: Research and practice*. New York: Cambridge.

Merriam, S. B. (Ed.). (1993, Spring). *An update on adult learning theory*. New Directions for Adult and Continuing Education, no. 57. San Francisco: Jossey-Bass.

Merriam, S. B. & Caffarella, R. S. (1991). *Learning in adulthood*. San Francisco: Jossey-Bass.

Miles, C. C. & Miles, W. R. (1932). The correlation of intelligence scores and chronological age from early to late maturity. *American Journal of Psychology, 44*, 44-78.

Miller, W. W., Doerfert, D. L., & King, J. C. (1996). Distributed education and interaction: The classroom and beyond. *Proceedings: Central Region Research Conference in Agricultural Education, 50*, 47-56.

Moore, M. G. (1989). Three types of interaction. *The American Journal of Distance Education, 3*(2), 1-6.

Moore, M. G., Thompson, M. M., Quiglely, B. A., Clark, G. C., & Goff, G. G. (1990). *The effect of distance learning: A summary of the literature*. Research Monograph No. 2. ERIC Document Reproduction Service No. 330 321).

Mott, V. W. & Rampp, L. C. (Eds.). (1995). *Organization and administration of continuing education*. Checotah, OK: AP Publications.

Naisbitt, J. (1982). *Megatrends: Ten new directions transforming our lives*. New York: Warner Books, Inc.

National Center for Education Statistics. (1993, June). *Adult education: Main reasons for participating*. Washington, DC: U.S. Department of Education, Office of Educational Research and Improvement.

National Center for Education Statistics. (1994, January). *Education and labor market outcomes of high school diploma and GED graduates*. Washington, DC: U.S. Department of Education, Office of Educational Research and Improvement.

National Center on Education and the Economy. (1990, June). *America's choice: High skills or low wages.* The Report of The Commission on the Skills of the American Workforce.

Newsome, R. (1992, November/December). Adult education needs a definition the public understands. *Adult Learning, 4*(2), 31.

Notar, E. E. (1994). *Solving the puzzle: Teaching and learning with adults.* New York: Rivercross Publishing, Inc.

Phillips, L. (1994). *The continuing education guide: The CEU and other professional education criteria.* Dubuque, IA: Kendall/Hunt.

Rachal, J. R. (1988). Taxonomies and typologies of adult education. *Lifelong learning: An omnibus of practice and research, 12*(2), 20-23.

Reece, B. L. (1985). *Teaching adults: A guide for instructors.* Arlington, VA: American Vocational Association.

Riley, R. W. (1993, July/August). Changing the culture: A lifetime of learning. *Adult Learning, 4*(6), 19-20.

Rosenbloom, S. H. (1985). *Involving adults in the educational process.* San Francisco: Jossey-Bass.

Scheneman, S. (1993, September/October). Continuing professional education: Education and learning. *Adult Learning, 5*(1), 6.

Schmidt, A. (1997). *A painless and simplified guide for preparing instructional visuals for electronic presentations.* Ames: Department of Agricultural Education and Studies, Iowa State University.

Seaman, D. F. & Fellenz, R. A. (1989). *Effective strategies for teaching adults.* Columbus, OH: Merrill Publishing Company, A Bell & Howell Information Company.

Simonson, M. (1995). Distance education revisited: An introduction to the issue. *Tech Trends, 40*(3), 2.

Simonson, M. & Schlosser, C. (1995). More than fiber: Distance education in Iowa. *Tech Trends, 40*(3), 13-15.

Smith, B. J. & Delahaye, B. L. (1987). *How to be an effective trainer.* New York: Wiley.

Stuart, R. & Holmes, L. (1982). Successful trainer styles. *Training and Development Journal, 6*(4), 17-23.

Stufflebeam, D. L., Foley, W. J., Gephart, W. J., Guba, E. G., Hammand, R. L., Merriam, H. O., & Provus, M. M. (1971). *Educational evaluation and decision making.* Itasca, IL: F. E. Peacock Publishers, Inc.

Thorndike, E. L., Bregman, E. O., Tilton, J. W. & Woodyard, E. (1928). *Adult Learning.* New York: The MacMillan Company.

Venable, W. R. & Mott, V. W. (1995). Needs assessment in continuing education. In V. W. Mott & L. C. Rampp (Eds.). *Organization & Administration of Continuing Education* (pp. 39-64). Checotah, OK: AP Publications.

Wagner, D. A. (1993, September/October). Myths and misconceptions in adult literacy. *Adult Learning, 5*(1), 9-10, 23.

Waldron, M. W. & Moore, G. A. B. (1991). *Helping adults learn: Course planning for adult learners.* Toronto: Thompson Educational Publishing, Inc.

Watkins, B. L. (1991). A quite radical idea: The invention and elaboration of collegiate correspondence study. In B. L. Watkins & S. J. Wright (Eds.). *The Foundations of American Distance Education.* (pp. 1-35). Dubuque, IA: Kendall/Hunt.

Willis, B. (1993). *Distance education: A practical guide.* Engelwood Cliffs, NJ: Educational Technology Publications.

Willis, B. (Ed.). (1994). *Distance Education Strategies and Tools* (p. v). Engelwood Cliffs, NJ: Educational Technology Publications.

Zahn, J. C. (1969, Spring/Summer). Differences between adults and youth affect learning. *Adult Jewish Education,* (27), 10-18.

Zemke, R. & Zemke, S. (1981). 30 things we know for sure about adult learning. *Training, 18,* 45-49.

INDEX